T0306083

Mathematics Manual for Water and Wastewater Treatment Plant Operators: Water Treatment Operations

To properly operate a waterworks or wastewater treatment plant and to pass the examination for a waterworks/wastewater operator's license, it is necessary to know how to perform certain calculations. All operators, at all levels of licensure, need a basic understanding of arithmetic and problem-solving techniques to solve the problems they typically encounter in the workplace.

Hailed on its first publication as a masterly account written in an engaging, highly readable, user-friendly style, the fully updated *Mathematics Manual for Water and Wastewater Treatment Plant Operators: Water Treatment Operations* covers all the necessary computations used in water treatment today. It presents math operations that progressively advance to higher, more practical applications, including math operations that operators at the highest level of licensure would be expected to know and perform.

Features:

- Provides a strong foundation based on theoretical math concepts, which it then applies to solving practical problems for both water and wastewater operations.
- Updated throughout and with several new practical problems added.
- Provides illustrative examples for commonly used waterworks and wastewater treatment operations covering unit process operations found in today's treatment facilities.

Mathematics Manual for Water and Wastewater Treatment Plant Operators: Water Treatment Operations

Math Concepts and Calculations

Third Edition

Frank R. Spellman

CRC Press
Taylor & Francis Group
Boca Raton London New York

CRC Press is an imprint of the
Taylor & Francis Group, an **informa** business

Third edition published 2024
by CRC Press
6000 Broken Sound Parkway NW, Suite 300, Boca Raton, FL 33487–2742

and by CRC Press
4 Park Square, Milton Park, Abingdon, Oxon, OX14 4RN

CRC Press is an imprint of Taylor & Francis Group, LLC

© 2024 Frank R. Spellman

First edition published by CRC Press 2004
Second edition published by CRC Press 2014

ISBN: 978-1-032-40687-9 (hbk)
ISBN: 978-1-032-40686-2 (pbk)
ISBN: 978-1-003-35430-7 (ebk)

DOI: 10.1201/9781003354307

Typeset in Times
by Apex CoVantage, LLC

Contents

Preface..xi
About the Author xiii

Chapter 1 Pumping Calculations 1
    1.1   Pumping...1
    1.2   Basic Water Hydraulics Calculations 1
        1.2.1   Weight of Water...1
        1.2.2   Weight of Water Related to the
                  Weight of Air 2
        1.2.3   Water at Rest 2
        1.2.4   Gauge Pressure 3
        1.2.5   Water in Motion...3
                1.2.5.1   Discharge 4
        1.2.6   Pipe Friction 5
    1.3   Basic Pumping Calculations..6
        1.3.1   Pumping Rates...6
        1.3.2   Calculating Head Loss..8
        1.3.3   Calculating Head 8
        1.3.4   Calculating Horsepower and Efficiency 9
                1.3.4.1   Hydraulic Horsepower (WHP) 9
                1.3.4.2   Pump Efficiency and Brake
                          Horsepower (BHP) 10
        1.3.5   Specific Speed 12
    1.4   Positive Displacement Pumps.. 13

Chapter 2 Water Source and Storage Calculations 15
    2.1   Water Sources 15
    2.2   Water Source Calculations... 16
        2.2.1   Well Drawdown... 16
        2.2.2   Well Yield 17
        2.2.3   Specific Yield... 18
        2.2.4   Well Casing Disinfection... 19
        2.2.5   Deep Well Turbine Pump Calculations 19
        2.2.6   Vertical Turbine Pump Calculations............................ 20
    2.3   Water Storage 25
    2.4   Water Storage Calculations 25
    2.5   Copper Sulfate Dosing... 27

2.6   Potassium Permanganate Dosing 30
       2.6.1   The 411 on Iron and Manganese Removal 30
               2.6.1.1   Iron and Manganese Removal
                         Techniques 30
               2.6.1.2   Precipitation 30
               2.6.1.3   Oxidation 31
               2.6.1.4   Ion Exchange 31
               2.6.1.5   Sequestering 32
               2.6.1.6   Aeration 32
               2.6.1.7   Potassium Permanganate Oxidation
                         and Manganese Greensand 32

Chapter 3   Coagulation and Flocculation Calculations 35
3.1   Coagulation 35
3.2   Flocculation 35
3.3   Coagulation and Flocculation Calculations 36
       3.3.1   Chamber and Basin Volume Calculations 36
       3.3.2   Detention Time 37
               3.3.2.1   Determining Dry Chemical Feeder
                         Setting (Lb/Day) 38
       3.3.3   Determining Chemical Solution Feeder
               Setting (GPD) 39
       3.3.4   Determining Chemical Solution Feeder
               Setting (ML/Min) 39
       3.3.5   Determining Percent of Solutions 40
       3.3.6   Determining Percent Strength of Liquid
               Solutions 41
       3.3.7   Determining Percent Strength of Mixed
               Solutions 42
       3.3.8   Dry Chemical Feeder Calibration 42
       3.3.9   Solution Chemical Feeder Calibration 43
       3.3.10 Determining Chemical Usage 45
3.4   Coagulation and Flocculation Practice Problems 46
3.5   Chemical Feeders 53
3.6   Chemical Feeder Calculations 54

Chapter 4   Sedimentation Calculations 59
4.1   Sedimentation 59
4.2   Tank Volume Calculations 59
       4.2.1   Calculating Tank Volume 59
4.3   Detention Time (DT) 60
4.4   Surface Overflow Rate (SOR) 61
4.5   Mean Flow Velocity 62
4.6   Weir Loading Rate (Weir Overflow Rate, WOR) 63

4.7   Percent Settled Biosolids..64
4.8   Determining Lime Dosage (Mg/L)..65
4.9   Determining Lime Dosage (Lb/Day)..68
4.10   Determining Lime Dosage (G/Min) 69
4.11   Sedimentation Practice Problems...70

Chapter 5   Filtration Calculations..79

5.1   Water Filtration...79
5.2   Flow Rate through a Filter (GPM) 79
5.3   Filtration Rate...82
5.4   Unit Filter Run Volume (UFRV)..84
5.5   Backwash Rate 85
       5.5.1   Backwash Rise Rate 87
5.6   Volume of Backwash Water Required (Gal)................................88
5.7   Required Depth of Backwash Water Tank (Ft)............................89
5.8   Backwash Pumping Rate (GPM)..89
5.9   Percent Product Water Used for Backwatering 90
5.10   Percent Mud Ball Volume..91
5.11   Filter Bed Expansion 92
5.12   Filter Loading Rate...93
5.13   Filter Medium Size 94
5.14   Mixed Media..95
5.15   Head Loss for Fixed-Bed Flow..96
5.16   Head Loss through a Fluidized Bed 97
5.17   Horizontal Washwater Troughs 99
5.18   Filter Efficiency 100
5.19   Water Filtration Practice Problems..101
5.20   Bibliography 106

Chapter 6   Water Chlorination Calculations.....................................109

6.1   Chlorine Disinfection 109
6.2   Determining Chlorine Dosage (Feed Rate) 109
6.3   Calculating Chlorine Dose, Demand, and Residual 111
6.4   Breakpoint Chlorination Calculations 113
6.5   Calculating Dry Hypochlorite Feed Rate 115
6.6   Calculating Hypochlorite Solution Feed Rate......................117
6.7   Calculating Percent Strength of Solutions............................119
       6.7.1   Calculating Percent Strength Using Dry
                Hypochlorite..119
       6.7.2   Calculating Percent Strength Using
                Liquid Hypochlorite 120
6.8   Chemical Use Calculations..120
6.9   Chlorination Math Practice Problems 121

Chapter 7   Fluoridation 125

          7.1   Water Fluoridation.. 125
          7.2   Fluoride Compounds 125
                 7.2.1   Sodium Fluoride 126
                 7.2.2   Fluorosilicic Acid 126
                 7.2.3   Sodium Fluorosilicate 127
          7.3   Optimal Fluoride Levels... 128
          7.4   Fluoridation Process Calculations 129
                 7.4.1   Percent Fluoride Ion in a Compound 129
                 7.4.2   Fluoride Feed Rate 130
                 7.4.3   Fluoride Feed Rates for Saturator 132
                 7.4.4   Calculated Dosages 133
                 7.4.5   Calculated Dosage Problems................................... 135
                 7.4.6   Fluoridation Math Practice Problems..................... 137

Chapter 8   Water Softening.. 141

          8.1   Water Hardness... 141
                 8.1.1   Calculating Calcium Hardness, as $CaCO_3$ 141
                 8.1.2   Calculating Magnesium Hardness, as $CaCO_3$ 142
                 8.1.3   Calculating Total Hardness..................................... 143
                 8.1.4   Calculating Carbonate and Noncarbonate
                           Hardness 144
          8.2   Alkalinity Determination.. 145
          8.3   Determining Bicarbonate, Carbonate, and Hydroxide
                 Alkalinity 146
          8.4   Lime Dosage Calculation for Removal of Carbonate
                 Hardness 148
          8.5   Calculation for Removal of Noncarbonate Hardness 150
          8.6   Recarbonation Calculation 151
          8.7   Calculating Feed Rates 152
          8.8   Ion Exchange Capacity 153
          8.9   Water Treatment Capacity 154
          8.10 Treatment Time Calculation (Until Regeneration
                 Required) 156
          8.11 Salt and Brine Required for Regeneration.............................. 156
          8.12 Salt Solutions Table 157

Chapter 9   Water Treatment Practice Calculations: Basic Math Problems........ 159

          9.1   Decimal Operations 159
          9.2   Percentage Calculations... 160
          9.3   Find x.. 161
          9.4   Ratio and Proportion... 162
          9.5   Area of Rectangles 163

Contents

9.6 Circumference and Area of Circles 163
9.7 Water Treatment Problems 164
9.8 Additional Practice Problems 234
    9.8.1 Tank Volume Calculations.......................................234
9.9 Channel and Pipeline Capacity Calculations.......................235
9.10 Miscellaneous Volume Calculations.....................................235
9.11 Flow, Velocity, and Conversion Calculations 237
9.12 Average Flow Rates...239
9.13 Flow Conversions 239
9.14 General Flow and Velocity Calculations 240
9.15 Chemical Dosage Calculations...242
9.16 Advanced Practice Problems..243

Appendix A: Basic Math Answers 253
Index..301

Preface

This is volume 2 of an industry-wide bestseller hailed on its first publication as a masterly account written in an engaging, highly readable, user-friendly, show-and-display—a classic present-and-do style—presentation. *Mathematics Manual for Water and Wastewater Treatment Plant Operators*, 3rd edition, volume 2 contains hundreds of worked examples presented in step-by-step training style; it is ideal for all levels of water treatment operators in training and practitioners studying for advanced licensure. In addition, this manual is a handy desk reference and/or hand-held guide for daily use in making operational math computations.

This standard synthesis not only has been completely revised but also expanded from one to three volumes. Volume 1 covers basic math operators and operations, volume 2 covers computations commonly used in water treatment plant operations, and volume 3 covers computations used in wastewater treatment.

To properly operate a waterworks or wastewater treatment plant and to pass the examination for a waterworks/wastewater operator's license, it is necessary to know how to perform certain calculations. In reality, most of the calculations that operators at the lower level of licensure need to know how to perform are not difficult, but all operators need a basic understanding of arithmetic and problem-solving techniques to be able to solve the problems they typically encounter.

How about waterworks/wastewater treatment plant operators at the higher levels of licensure? Do they also need to be well-versed in mathematical operations? The short answer is absolutely. The long answer is that if you work in water or wastewater treatment and expect to have a successful career which includes advancement to the highest levels of licensure or certification (usually prerequisites for advancement to higher management levels), you must have knowledge of math at both the basic or fundamental level and advanced practical level. It is not possible to succeed in this field without the ability to perform mathematical operations.

Keep in mind that mathematics is a universal language. Mathematical symbols have the same meaning to people speaking many different languages throughout the world. The key to learning mathematics is to learn the language, symbols, definitions, and terms of mathematics that allow us to grasp the concepts necessary to solve equations.

In *Mathematics Manual for Water/Wastewater Treatment Plant Operators*, we begin by introducing and reviewing concepts critical to the qualified operators at the fundamental or entry level; however, this does not mean that these are the only math concepts that a competent operator needs to know to solve routine operation and maintenance problems. After covering the basics in volume 1, volumes 2 and 3 present industry-wide math operations that progressively advance, step-by-step, to higher, more practical applications of mathematical calculations—that is, the math operations that operators at the highest level of licensure would be expected to know how to perform.

After building a strong foundation based on theoretical math concepts (the basic tools of mathematics, including fractions, decimals, percent, areas, volumes) in volume 1, we move on to applied math presented in this volume. Even though there is considerable crossover of basic math operations used by both waterworks and wastewater operators, we separate applied math problems for wastewater and water. We do this to aid operators of specific unit processes unique to waterworks and wastewater operations focus on their area of specialty.

What makes *Mathematics Manual for Water/Wastewater Treatment Plant Operators Volume 2* different from the other available math books available? Consider the following:

- The author has worked in and around water/wastewater treatment and taught water/wastewater math for several years at the apprenticeship level and at numerous short courses for operators.
- The author has sat at the table of licensure examination preparation boards to review, edit, and write state licensure exams.
- This step-by-step training manual provides concise, practical instruction in the math skills that operators must have to pass certification tests.
- The text is completely self-contained in three complete volumes. The advantage should be obvious—three separate texts with math basics and advanced operator math concepts contained in each allow the user to choose the proper volume for his or her use.
- The text is user-friendly; no matter the difficulty of the problem to be solved, each operation is explained in straightforward, plain English. Moreover, numerous example problems (several hundred) are presented to enhance the learning process.
- The first edition was highly successful and well-received, but like any flagship edition of any practical manual, there is always room for improvement. Thankfully, many users have provided constructive criticism, advices, and numerous suggestions. All these inputs from actual users have been incorporated into this new three-volume set.

To assure correlation to modern practice and design, we present illustrative problems in terms of commonly used waterworks/wastewater treatment operations and associated parameters and cover typical math concepts for waterworks/wastewater treatment unit process operations found in today's waterworks/wastewater treatment facilities.

This text is accessible to those who have little or no experience in treatment plant math operations. Readers who work through the text systematically will be surprised at how easily they can acquire an understanding and skill in water/wastewater math concepts, adding another critical component to your professional knowledge.

A final point before beginning our discussion of math concepts: it can be said with some accuracy and certainty that without the ability to work basic math problems (i.e., those typical to water/wastewater treatment), candidates for licensure will find any attempts to successfully pass licensure exams a much more difficult proposition.

Frank R. Spellman
Norfolk, VA

About the Author

Frank R. Spellman is a retired assistant professor of environmental health at Old Dominion University, Norfolk, Virginia, and the author of more than 160 books covering topics ranging from concentrated animal feeding operations (CAFOs) to all areas of environmental science and occupational health. He consults on homeland security vulnerability assessments for critical infrastructures, including water/ wastewater facilities, and conducts audits for Occupational Safety and Health Administration and Environmental Protection Agency inspections throughout the country. Dr. Spellman lectures on sewage treatment, water treatment, and homeland security, as well as on safety topics, throughout the country and teaches water/wastewater operator short courses at Virginia Tech in Blacksburg.

1 Pumping Calculations

1.1 PUMPING

Pumps and pumping calculations were discussed in detail in volume 1 and are discussed in this volume and in volume 3 because they are germane to many treatment processes and especially to their influent and effluent operations. Pumping facilities and appurtenances are required wherever gravity can't be used to supply water to the distribution system under sufficient pressure to meet all service demands. Pumps used in water and wastewater treatment are the same. Because the pump is so perfectly suited to the tasks it performs, and because the principles that make the pump work are physically fundamental, the idea that any new device would ever replace the pump is difficult to imagine. The pump is the workhorse of water/wastewater operations. Simply, pumps use energy to keep water and wastewater moving. To operate a pump efficiently, the operator and/or maintenance operator must be familiar with several basic principles of hydraulics. In addition, to operate various unit processes, in both water and wastewater operations at optimum levels, operators should know how to perform basic pumping calculations.

1.2 BASIC WATER HYDRAULICS CALCULATIONS

1.2.1 WEIGHT OF WATER

Because water must be stored and/or kept moving in water supplies and wastewater must be collected, processed, and discharged (out-falled) to its receiving body, we must consider some basic relationships in the weight of water; 1 cu ft of water weighs 62.4 lb and contains 7.48 gal, while 1 cu in of water weighs 0.0362 lb. Water 1 ft deep will exert a pressure of 0.43 psi on the bottom area (12 in $\times$ 0.062 lb/in^3). A column of water 2 ft high exerts 0.86 psi, one 10 ft high exerts 4.3 psi, and one 52 ft high exerts:

$$52\,\text{ft} \times 0.43\,\text{psi/ft} = 22.36\,\text{psi}$$

A column of water 2.31 ft high will exert 1.0 psi. To produce a pressure of 40 psi requires a water column of:

$$40\,\text{psi} \times 2.31\,\text{ft/psi} = 92.4$$

The term *head* is used to designate water pressure in terms of the height of a column of water in feet. For example, a 10 ft column of water exerts 4.3 psi. This can be called 4.3 psi pressure or 10 ft of head. If the static pressure in a pipe leading from an elevated water storage tank is 37 psi (pounds per square inch), what is the elevation of the water above the pressure gauge?

Remembering that 1 psi = 2.31 and that the pressure at the gauge is 37 psi:

$$37\,\text{psi} \times 2.31\,\text{ft/psi} = 85.5\,\text{ft}\,(\text{rounded})$$

DOI: 10.1201/9781003354307-1

1.2.2   WEIGHT OF WATER RELATED TO THE WEIGHT OF AIR

The theoretical atmospheric pressure at sea level (14.7 psi) will support a column of water 34 ft high:

$$14.7\,psi \times 2.31\,ft/psi = 33.957\,ft \text{ or } 34 \text{ ft}$$

At an elevation of 1 mi above sea level, where the atmospheric pressure is 12 psi, the column of water would be only 28 ft high:

$$\left(12\,psi \times 2.31\,ft/psi = 27.72\,ft \text{ or } 28\,ft\right).$$

    If a tube is placed in a body of water at sea level (a glass, a bucket, a water storage reservoir, or a lake, pool, etc.), water will rise in the tube to the same height as the water outside the tube. The atmospheric pressure of 14.7 psi will push down equally on the water surface inside and outside the tube. However, if the top of the tube is tightly capped and all the air is removed from the sealed tube above the water surface, forming a *perfect vacuum*, the pressure on the water surface inside the tube will be 0 psi. The atmospheric pressure of 14.7 psi on the outside of the tube will push the water up into the tube until the weight of the water exerts the same 14.7 psi pressure at a point in the tube even with the water surface outside the tube. The water will rise 14.7 psi × 2.31 ft/psi = 34 ft.

    In practice, it is impossible to create a perfect vacuum, so the water will rise somewhat less than 34 ft; the distance it rises depends on the amount of vacuum created.

Example 1.1

Problem:

If enough air was removed from the tube to produce an air pressure of 9.5 psi above the water in the tube, how far will the water rise in the tube?

Solution:

To maintain the 14.7 psi at the outside water surface level, the water in the tube must produce a pressure of 14.7 psi – 9.5 psi = 5.2 psi. The height of the column of water that will produce 5.2 psi is:

$$5.2\,psi \times 2.31\,ft/psi = 12\,ft$$

1.2.3   WATER AT REST

Stevin's law states, "The pressure at any point in a fluid at rest depends on the distance measured vertically to the free surface and the density of the fluid." Stated as a formula, this becomes:

$$p = w \times h \tag{1.1}$$

Where:
    p = pressure in pounds per square foot (psf)
    w = density in pounds per cubic foot (lb/ft³)
    h = vertical distance in feet

Example 1.2

Problem:

What is the pressure at a point 16 ft below the surface of a reservoir?

Solution:

To calculate this, we must know that the density of water, w, is 62.4 lb/cu ft. Thus:

$$p = w \times h$$
$$= 62.4 \, lb/ft^3 \times 16 \, ft$$
$$= 998.4 \, lb/ft^2 \text{ or } 998.4 \, psf$$

Waterworks operators generally measure pressure in pounds per square **inch** rather than pounds per square **foot**; to convert, divide by 144 in²/ft² (12 in × 12 in = 144 in²):

$$p = \frac{998.4 \, lb/ft^2}{144 \, in^2/ft^2} = 6.93 \, lb/in^2 \text{ or psi}$$

1.2.4 GAUGE PRESSURE

We have defined *head* as the height a column of water would rise due to the pressure at its base. We demonstrated that a perfect vacuum plus atmospheric pressure of 14.7 psi would lift the water 34 ft. If we now open the top of the sealed tube to the atmosphere and enclose the reservoir, then increase the pressure in the reservoir, the water will again rise in the tube. Because atmospheric pressure is essentially universal, we usually ignore the first 14.7 psi of actual pressure measurements and measure only the difference between the water pressure and the atmospheric pressure; we call this *gauge pressure*.

Example 1.3

Problem:

Water in an open reservoir is subjected to the 14.7 psi of atmospheric pressure, but subtracting this 14.7 psi leaves a gauge pressure of 0 psi. This shows that the water would rise 0 ft above the reservoir surface. If the gauge pressure in a water main is 110 psi, how far would the water rise in a tube connected to the main?

Solution:

$$110 \, psi \times 2.31 \, ft/psi = 254.1 \, ft$$

1.2.5 WATER IN MOTION

The study of water in motion is much more complicated than that of water at rest. It is important to have an understanding of these principles because the water/wastewater in a treatment plant and/or distribution/collection system is nearly always in motion (much of this motion is the result of pumping, of course).

1.2.5.1 Discharge

Discharge is the quantity of water passing a given point in a pipe or channel during a given period of time. It can be calculated by the formula:

$$Q = V \times A \tag{1.2}$$

Where:
    Q = discharge in cubic feet per second (cfs)
    V = water velocity in feet per second (fps or ft/sec)
    A = cross-section area of the pipe or channel in square feet (ft²)

Discharge can be converted from cubic feet per second to other units, such as gallons per minute (gpm) or million gallons per day (MGD), by using appropriate conversion factors.

Example 1.4

Problem:

A pipe 12 in in diameter has water flowing through it at 12 ft/sec. What is the discharge in (a) cubic feet per second, (b) gallons per minute, and (c) million gallons per day?

Solution:

Before we can use the basic formula, we must determine the area (A) of the pipe. The formula for the area is:

$$A = \pi \times \frac{D^2}{4} = \pi \times r^2$$

(π is the constant value 3.14159)

Where:
    D = diameter of the circle in feet
    r = radius of the circle in feet

So the area of the pipe is:

$$A = \pi \times \frac{D^2}{4} = 3.14159 = 0.785 \, \text{ft}^2$$

Now we can determine the discharge in cubic feet per second (part [a]):

$$Q = V \times A = 12 \, \text{ft/sec} \times 0.785 \, \text{ft}^2 = 9.42 \, \text{ft}^3/\text{sec or cfs}$$

For part (b), we need to know that 1 cfs is 449 gpm, so 7.85 cfs × 449 gpm/cfs = 3,520 gpm.

    Finally, for part (c), 1 MGD is 1.55 cfs, so:

$$\frac{7.85 \, \text{cfs}}{1.55 \, \text{cfs/MGD}} = 5.06 \, \text{MGD}$$

THE LAW OF CONTINUITY

The *law of continuity* states that the discharge at each point in a pipe or channel is the same as the discharge at any other point (provided water does not leave or enter the pipe or channel). In equation form, this becomes:

$$Q_1 = Q_2 \text{ or } A_1 V_1 = A_2 V_2 \qquad (1.3)$$

Example 1.5

Problem:

A pipe 12 in in diameter is connected to a 6 in diameter pipe. The velocity of the water in the 12 in pipe is 3 fps. What is the velocity in the 6 in pipe? Using the equation $A_1 V_1 = A_2 V_2$, we need to determine the area of each pipe.

$$12 \text{ inch pipe} : A = \pi \times \frac{D^2}{4}$$

$$= 3.1419 \times \frac{(1\,\text{ft})^2}{4}$$

$$= 0.785\,\text{ft}^2$$

$$6 \text{ inch pipe} : A = 3.14159 \times \frac{(0.5)^2}{4}$$

$$= 0.196\,\text{ft}^2$$

The continuity equation now becomes:

$$(0.785 \text{ ft}^2) \times (3 \text{ ft/sec}) = (0.196 \text{ ft}^2) \times V_2$$

Solving for V_2:

$$V_2 = \frac{(0.785\,\text{ft}^2) \times 3\,\text{ft/sec}}{(0.196\,\text{ft}^2)}$$

$$= 12 \text{ ft/sec or fps}$$

1.2.6 PIPE FRICTION

The flow of water in pipes is caused by the pressure applied behind it either by gravity or by hydraulic machines (pumps). The flow is retarded by the friction of the water against the inside of the pipe. The resistance of flow offered by this friction depends on the size (diameter) of the pipe, the roughness of the pipe wall, and the number and type of fittings (bends, valves, etc.) along the pipe. It also depends on the speed of the water through the pipe—the more water you try to pump through a pipe, the more pressure it will take to overcome the friction. The resistance can be expressed in terms of the additional pressure needed to push the water through the pipe, in either pounds per square inch or feet of head. Because it is a reduction in pressure, it is often referred to as *friction loss* or *head loss*.

FRICTION LOSS INCREASES AS:

- Flow rate increases
- Pipe diameter decreases
- Pipe interior becomes rougher
- Pipe length increases
- Pipe is constricted
- Bends, fittings, and valves are added

The actual calculation of friction loss is beyond the scope of this text. Many published tables give the friction loss in different types and diameters of pipe and standard fittings. What is more important here is recognition of the loss of pressure or head due to the friction of water flowing through a pipe. One of the factors in friction loss is the roughness of the pipe wall. A number called the C factor indicates pipe wall roughness; the **higher** the C factor, the **smoother** the pipe.

Note: C factor is derived from the letter C in the Hazen–Williams equation for calculating water flow through a pipe.

Some of the roughness in the pipe will be due to the material—cast iron pipe will be rougher than plastic, for example. Additionally, the roughness will increase with corrosion of the pipe material and deposits of sediments in the pipe. New water pipes should have a C factor of 100 or more; older pipes can have C factors very much lower than this. To determine C factor, we usually use published tables. In addition, when the friction losses for fittings are factored in, other published tables are available to make the proper determinations. It is standard practice to calculate the head loss from fittings by substituting the *equivalent length of pipe*, which is also available from published tables.

1.3 BASIC PUMPING CALCULATIONS

Certain computations used for determining various pumping parameters are important to the water/wastewater operator. In this section, we cover those basic pumping calculations important to the subject matter.

1.3.1 PUMPING RATES

Important Point: The rate of flow produced by a pump is expressed as the volume of water pumped during a given period.

The mathematical problems most often encountered by water/wastewater operators in regard to determining pumping rates are often determined by using equations 1.4 and/or 1.5.

$$\text{Pumping Rate, (gpm)} = \frac{\text{gallons}}{\text{minutes}} \tag{1.4}$$

$$\textit{Pumping Rate, } (gph) = \frac{\text{gallons}}{\text{hours}} \tag{1.5}$$

Example 1.6

Problem:

The meter on the discharge side of the pump reads in hundreds of gallons. If the meter shows a reading of 110 at 2:00 p.m. and 320 at 2:30 p.m., what is the pumping rate expressed in gallons per minute?

Solution:

The problem asks for pumping rate in gallons per minute (gpm), so we use equation 14.4.

$$\text{Pumping Rate, gpm} = \frac{\text{gallons}}{\text{minutes}}$$

Step 1: To solve this problem, we must first find the total gallons pumped (determined from the meter readings).

$$\begin{array}{r} 32,000 \text{ gallons} \\ -11,000 \text{ gallons} \\ \hline 21,000 \text{ gallons} \end{array}$$

Step 2: The volume was pumped between 2:00 p.m. and 2:30 p.m., for a total of 30 min. From this information, calculate the gallons per minute pumping rate:

$$\text{Pumping rate, gpm} = \frac{21,000 \text{ gal}}{30 \text{ min}}$$
$$= 700 \text{ gpm pumping rate}$$

Example 1.7

Problem:

During a 15 min pumping test, 16,000 gal were pumped into an empty rectangular tank. What is the pumping rate in gallons per minute?

Solution:

The problem asks for the pumping rate in gallons per minute, so again we use equation 1.4.

$$\text{Pumping rate, gpm} = \frac{\text{gallons}}{\text{minutes}}$$
$$= \frac{16,000 \text{ gallons}}{15 \text{ minutes}}$$
$$= 1,067 \text{ gpm pumping rate}$$

Example 1.8

Problem:

A tank 50 ft in diameter is filled with water to a depth of 4 ft. To conduct a pumping test, the outlet valve to the tank is closed and the pump is allowed to discharge into the tank. After 60 min, the water level is 5.5 ft. What is the pumping rate in gallons per minute?

Solution:

Step 1: We must first determine the volume pumped in cubic feet:

$$\text{Volume pumped} = (\text{area of circle})(\text{depth})$$
$$= (0.785)\,(50\,\text{ft})\,(50\,\text{ft})\,(1.5\,\text{ft})$$
$$= 2,944\,\text{ft}^3 \ \ (\text{rounded})$$

Step 2: Convert the cubic feet volume to gallons:

$$(2,944\ \text{ft}^3)\,(7.48\ \text{gal/ft}^3) = 22,021\ \text{gallons (rounded)}$$

Step 3: The pumping test was conducted over a period of 60 min. Using equation 1.4, calculate the pumping rate in gallons per minute:

$$\text{pumping rate} = \frac{\text{gallons}}{\text{minutes}}$$
$$= \frac{22,021\,\text{gallons}}{60\,\text{minutes}}$$
$$= 267\,\text{gpm} \ \left(\text{rounded}\right)$$

1.3.2 Calculating Head Loss

Important Note: Pump head measurements are used to determine the amount of energy a pump can or must impart to the water; they are measured in feet.

One of the principal calculations in pumping problems is used to determine *head loss*. The following formula is used to calculate head loss:

$$H_f = K(V^2/2g) \tag{1.6}$$

Where:
H_f = friction head
K = friction coefficient
V = velocity in pipe
g = gravity (32.17 ft/sec)

1.3.3 Calculating Head

For centrifugal pumps and positive displacement pumps, several other important formulas are used in determining *head*. In centrifugal pump calculations, the conversion of the discharge pressure to discharge head is the norm. Positive displacement pump calculations often leave given pressures in pounds per square inch. In the following formulas, *W* expresses the specific weight of liquid in pounds per cubic foot. For water at 68°F, W is 62.4 lb/ft³. A water column 2.31 ft high exerts a pressure of 1 psi on 64°F water. Use the following formulas to convert discharge pressure in pounds per square gauge to head in feet:

- Centrifugal pumps

$$H, ft = \frac{P, psig \times 2.31}{specific\ gravity} \qquad (1.7)$$

- Positive displacement pumps

$$H, ft = \frac{P, psig \times 144}{W} \qquad (1.8)$$

To convert head into pressure:

- Centrifugal pumps

$$P, psi = \frac{H, ft \times specific\ gravity}{2.31} \qquad (1.9)$$

- Positive displacement pumps

$$P, psi = \frac{H, ft \times W}{W} \qquad (1.10)$$

1.3.4   CALCULATING HORSEPOWER AND EFFICIENCY

When considering work being done, we consider the "rate" at which work is being done. This is called *power* and is labeled as **foot-pounds per second**. At some point in the past, it was determined that the ideal work animal, the horse, could move 550 lb 1 ft in 1 sec. Because large amounts of work are also to be considered, this unit became known as *horsepower*. When pushing a certain amount of water at a given pressure, the pump performs work; 1 hp equals 33,000 ft-lb/min. The two basic terms for horsepower are:

- Hydraulic horsepower (whp)
- Brake horsepower (bhp)

1.3.4.1   Hydraulic Horsepower (WHP)

One hydraulic horsepower equals the following:

- 550 ft-lb/sec
- 33,000 ft lb/min
- 2,545 Btu/hr (British thermal units per hour)
- 0.746 kW
- 1.014 metric hp

To calculate the hydraulic horsepower (whp) using flow in gallons per minute and head in feet, use the following formula for centrifugal pumps:

$$\text{WHP} = \frac{\text{flow, gpm} \times \text{hear, ft} \times \text{specific gravity}}{3,960} \qquad (1.11)$$

When calculating horsepower for positive displacement pumps, common practice is to use pounds per square inch for pressure. Then the hydraulic horsepower becomes:

$$\text{WHP} = \frac{\text{flow, gpm} \times \text{pressure, psi}}{3,960} \qquad (1.12)$$

1.3.4.2 Pump Efficiency and Brake Horsepower (BHP)

When a motor–pump combination is used (for any purpose), neither the pump nor the motor will be 100% efficient. Simply, not all the power supplied by the motor to the pump (called *brake horsepower*, bhp) will be used to lift the water (*water or hydraulic horsepower*)—some of the power is used to overcome friction within the pump. Similarly, not all the power of the electric current driving the motor (called *motor horsepower*, mhp) will be used to drive the pump—some of the current is used to overcome friction within the motor, and some current is lost in the conversion of electrical energy to mechanical power.

Note: Depending on size and type, pumps are usually 50–85% efficient, and motors are usually 80–95% efficient. The efficiency of a particular motor or pump is given in the manufacturer's technical manual accompanying the unit.

The pump's *brake horsepower (bhp)* is equal to hydraulic horsepower divided by the pump's efficiency. Thus, the brake horsepower formula becomes:

$$\text{bhp} = \frac{\text{flow, gpm} \times \text{head, ft} \times \text{specific gravity}}{3,960 \times \text{efficiency}} \qquad (1.13)$$

or

$$\text{bhp} = \frac{\text{flow, gpm} \times \text{pressure, psig}}{1,714 \times \text{efficiency}} \qquad (1.14)$$

Example 1.9

Problem:

Calculate the brake horsepower requirements for a pump handling salt water and having a flow of 700 gpm with 40 psi differential pressure. The specific gravity of salt water at 68°F equals 1.03. The pump efficiency is 85%.

Solution:

To use equation 1.13, convert the pressure differential to total differential head, TDH = 40 × 2.31/1.03 = 90 ft.

$$\text{bhp} = \frac{700 \times 90 \times 1.03}{3,960 \times 0.85}$$

$$= 19.3 \text{ hp (rounded)}$$

Use equation 1.14:

$$bhp = \frac{700 \times 40}{1,714 \times 0.85}$$

$$= 19.2 \text{ hp (rounded)}$$

Important Point: Horsepower requirements vary with flow. Generally, if the flow is greater, the horsepower required to move the water would be greater.

When the motor, brake, and motor horsepower are known and the **efficiency** is unknown, a calculation to determine motor or pump efficiency must be done. Equation 14.15 is used to determine percent efficiency.

$$\text{Percent Efficiency} = \frac{\text{hp output}}{\text{hp input}} \times 100 \qquad (1.15)$$

From equation 1.15, the specific equations to be used for motor, pump, and overall efficiency equations are:

$$\text{Percent Motor Efficiency} = \frac{\text{bhp}}{\text{mhp}} \times 100$$

$$\text{Percent Pump Efficiency} = \frac{\text{whp}}{\text{bhp}} \times 100$$

$$\text{Percent Overall Efficiency} = \frac{\text{whp}}{\text{mhp}} \times 100$$

Example 1.10

Problem:

A pump has a water horsepower requirement of 8.5 whp. If the motor supplies the pump with 10 hp, what is the efficiency of the pump?

Solution:

$$\text{Percent pump efficiency} = \frac{\text{whp output}}{\text{bhp supplied}} \times 100$$

$$= \frac{8.5 \text{ whp}}{10 \text{ bhp}} \times 100$$

$$= 0.85 \times 100$$

$$= 85\%$$

Example 1.11

Problem:

What is the efficiency if an electric power equivalent to 25 hp is supplied to the motor and 16 hp of work is accomplished by the pump?

Solution:

Calculate the percent of overall efficiency:

$$\text{Percent overall efficiency} = \frac{\text{hp output}}{\text{hp supplied}} \times 100$$

$$= \frac{16 \text{ whp}}{25 \text{ mhp}} \times 100$$

$$= 0.64 \times 100$$

$$= 64\%$$

Example 1.12

Problem:

12 kW (kilowatts) of power is supplied to the motor. If the brake horsepower is 12 hp, what is the efficiency of the motor?

Solution:

First, convert the kilowatts power to horsepower. Based on the fact that 1 hp = 0.746 Kw, the equation becomes:

$$\frac{12 \text{ kW}}{0.746 \text{ kW/hp}} = 16.09 \text{ hp}$$

Now, calculate the percent efficiency of the motor:

$$\text{Percent efficiency} = \frac{\text{hp output}}{\text{hp supplied}} \times 100$$

$$= \frac{12 \text{ bhp}}{16.09 \text{ mhp}} \times 100$$

$$= 75\%$$

1.3.5 SPECIFIC SPEED

Specific speed (N_s) refers to an impeller's speed when pumping 1 gpm of liquid at a differential head of 1 ft. Use the following equation for specific speed, where H is at the best efficiency point:

$$N_s = \frac{\text{rpm} \times Q^{0.5}}{H^{0.75}} \qquad (1.16)$$

Where:
  rpm = revolutions per minute
  Q = flow (in gallons per minute)
  H = head (in feet)

Pump specific speeds vary between pumps. No absolute rule sets the specific speed for different kinds of centrifugal pumps. However, the following N_s ranges are quite common.

- Volute, diffuser, and vertical turbine: 500–5,000
- Mixed flow: 5,000–10,000
- Propeller pumps: 9,000–15,000

Important Note: The higher the specific speed of a pump, the higher its efficiency.

1.4 POSITIVE DISPLACEMENT PUMPS

The clearest differentiation between centrifugal (or kinetic) pumps and positive displacement pumps is made or based on the method by which pumping energy is transmitted to the liquid. Kinetic (centrifugal pumps) relies on a transformation of kinetic energy to static pressure. Positive displacement pumps, on the other hand, discharge a given volume for each stroke or revolution (i.e., energy is added intermittently to the fluid flow). The two most common forms of positive displacement pumps are reciprocating action pumps (which use pistons, plungers, diaphragms, or bellows) and rotary action pumps (using vanes, screws, lobes, or progressing cavities). Regardless of form used, all positive displacement pumps act to force liquid into a system regardless of the resistance that may oppose the transfer. The discharge pressure generated by a positive displacement pump is, in theory, infinite.

THE THREE BASIC TYPES OF POSITIVE DISPLACEMENT PUMPS DISCUSSED IN THIS CHAPTER ARE:

- Reciprocating pumps
- Rotary pumps
- Special-purpose pumps (peristaltic or tubing pumps)

Volume of Biosolids Pumped (Capacity)

One of the most common positive displacement biosolids pumps is the piston pump. Each stroke of a piston pump "displaces" or pushes out biosolids. Normally, the piston pump is operated at about 50 gpm. In making capacity for positive displacement pump calculations, we use the volume of biosolids pumped equation shown in the following.

$$\text{Vol. of Biosolids Pumped, gal/min} = [(0.785)\,(D^2)\,(\text{Stroke Length}) \tag{1.17}$$
$$(7.48\ \text{gal/cu ft}]\,[\text{No. of Strokes/min}]$$

Example 1.13

Problem:

A biosolids pump has a bore of 6 in and a stroke length of 4 in. If the pump operates at 55 strokes (or revolutions) per minute, how many gallons per minute are pumped? (Assume 100% efficiency.)

Solution:

> Vol. of Biosolids Pumped = (Gallons pumped/stroke) (No. of Strokes/minute)
> $= [(0.785)(D^2)(\text{Stroke Length}) (7.48 \text{ gal/cu ft})]$ [Strokes/min]
> $= [(0.785) (0.5 \text{ ft}) (0.5 \text{ ft}) (0.33 \text{ ft}) (7.48 \text{ gal/cu ft})]$ [55 strokes/min]
> $= (0.48 \text{ gal/stroke}) (50 \text{ strokes/min})$
> $= 26.6$ gpm

Example 1.14

Problem:

A biosolids pump has a bore of 6 in and a stroke setting of 3 in. The pump operates at 50 rpm. If the pump operates a total of 60 min during a 24 hr period, what is the gallons per day pumping rate? (Assume the piston is 100% efficient.)

First, calculate the gallons per minute pumping rate:

> Vol. Pumped, gpm
> $\qquad = (\text{Gallons pumped/Stroke}) (\text{No. of Strokes/minute})$
> $\qquad = [(0.785) (0.5 \text{ ft}) (0.5 \text{ ft}) (0.25 \text{ ft}) (7.48 \text{ gal/cu ft})]$ [50 Strokes/min]
> $\qquad = (0.37 \text{ gal/stroke}) (50 \text{ strokes/min})$
> $\qquad = 18.5$ gpm

Then convert gallons per minute to gallons per day pumping rate, based on total minutes pumped during 24 hr:

$$= (18.5) (60/\text{day})$$
$$= 1110 \text{ gpd}$$

2 Water Source and Storage Calculations

2.1 WATER SOURCES

Approximately 40 million cubic miles of water cover or reside within the Earth. The oceans contain about 97% of all water on Earth. The other 3% is fresh water: (1) snow and ice on the surface of Earth contain about 2.25% of the water, (2) usable groundwater is approximately 0.3%, and (3) surface freshwater is less than 0.5%. In the United States, for example, average rainfall is approximately 2.6 ft (a volume of 5,900 cu km). Of this amount, approximately 71% evaporates (about 4,200 cu km), and 29% goes to stream flow (about 1,700 cu km).

Beneficial freshwater uses include manufacturing, food production, domestic and public needs, recreation, hydroelectric power production, and flood control. Stream flow withdrawn annually is about 7.5% (440 cu km). Irrigation and industry use almost half of this amount (3.4%, or 200 cu km/year). Municipalities use only about 0.6% (35 cu km/year) of this amount.

Historically, in the United States, water usage is increasing (as might be expected). For example, in 1900, 40 billion gallons of freshwater were used. In 1975, the total increased to 455 billion gallons. Projected use in 2000 is about 720 billion gallons.

The primary sources of fresh water include the following:

- Captured and stored rainfall in cisterns and water jars
- Groundwater from springs, artesian wells, and drilled or dug wells
- Surface water from lakes, rivers, and streams
- Desalinized seawater or brackish groundwater
- Reclaimed wastewater

In addition to water source calculations that are typically used by water treatment operators in the operation of waterworks reservoirs, storage ponds, and lakes and are also included in many operator certification examinations, additional example math problems on algicide application have been included in this chapter. State public water supply divisions issue permits to allow the application of copper sulfate and potassium permanganate to water supply reservoirs for algae control. Moreover, iron and manganese control via chemical treatment is, in many cases, also required. Thus, water operators must be trained in the proper pretreatment (proper dosage) of water storage reservoirs, ponds, and/or lakes for control of algae, iron, and manganese.

DOI: 10.1201/9781003354307-2

2.2 WATER SOURCE CALCULATIONS

Water source calculations covered in this section apply to wells and pond/lake storage capacity. Specific well calculations discussed include well drawdown, well yield, specific yield, well casing disinfection, and deep well turbine pump capacity.

2.2.1 WELL DRAWDOWN

Drawdown is the drop in the level of water in a well when water is being pumped (see Figure 2.1). Drawdown is usually measured in feet or meters. One of the most important reasons for measuring drawdown is to make sure that the source water is adequate and not being depleted. The data that is collected to calculate drawdown can indicate if the water supply is slowly declining. Early detection can give the system time to explore alternative sources, establish conservation measures, or obtain any special funding that may be needed to get a new water source. Well drawdown is the difference between the pumping water level and the static water level.

$$\text{Drawdown, ft} = \text{Pumping Water Level, ft} - \text{Static Water Level, ft} \qquad (2.1)$$

Example 2.1

Problem:

The static water level for a well is 70 ft. If the pumping water level is 110 ft, what is the drawdown?

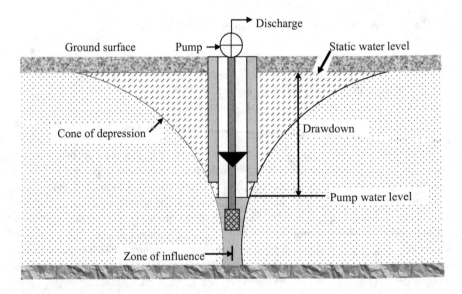

FIGURE 2.1 Drawdown.

Solution:

$$\text{Drawdown, ft} = \text{Pumping Water Level, ft} - \text{Static Water Level, ft}$$
$$= 110 \text{ ft} - 70 \text{ ft}$$
$$= 40 \text{ ft}$$

Example 2.2

Problem:

The static water level of a well is 125 ft. The pumping water level is determined using the sounding line. The air pressure applied to the sounding line is 4 psi, and the length of the sounding line is 180 ft. What is the drawdown?

Solution:

First, calculate the water depth in the sounding line and the pumping water level:

1. Water depth in sounding line = (4.0 psi) (2.31 ft/psi)
   = 9.2 ft
2. Pumping water level = 180 ft – 9.2 ft = 170.8 ft.

Then calculate drawdown as usual:

$$\text{Drawdown, ft} = \text{Pumping Water Level, ft} - \text{Static Water Level, ft}$$
$$= 170.8 \text{ ft} - 125 \text{ ft}$$
$$= 45.8 \text{ ft}$$

2.2.2 WELL YIELD

Well yield is the volume of water per unit of time that is produced from the well pumping. Usually, well yield is measured in terms of gallons per minute (gpm) or gallons per hour (gph). Sometimes, large flows are measured in cubic feet per second (cfs). Well yield is determined by using the following equation:

$$\text{Well Yield, gpm} = \frac{\text{Gallons Produced}}{\text{Duration of Test, min}} \qquad (2.2)$$

Example 2.3

Problem:

Once the drawdown level of a well stabilized, it was determined that the well produced 410 gal during a 5 min test.

Solution:

$$\text{Well Yield, gpm} = \frac{\text{Gallons Produced}}{\text{Duration of Test, min}}$$

$$= \frac{410 \text{ gallons}}{5 \text{ minutes}}$$
$$= 82 \text{ gpm}$$

Example 2.4

Problem:

During a 5 min test for well yield, a total of 760 gal is removed from the well. What is the well yield in gallons per minute? In gallons per hour?

Solution:

$$\text{Well Yield, gpm} = \frac{\text{Gallons Removed}}{\text{Duration of Test, min}}$$
$$= \frac{760 \text{ gallons}}{5\text{-minutes}}$$
$$= 152 \text{ gpm}$$

Then convert gallons per minute flow to gallons per hour flow:

$$(152 \text{ gal/min}) \ (60/\text{hr}) = 9120 \text{ gph})$$

2.2.3 SPECIFIC YIELD

Specific yield is the discharge capacity of the well per foot of drawdown. The specific yield may range from 1 gpm/ft drawdown to more than 100 gpm/ft drawdown for a properly developed well. Specific yield is calculated using the equation:

$$\text{Specific Yield, gpm/ft} = \frac{\text{Well Yield, gpm}}{\text{Drawdown, ft}} \qquad (2.3)$$

Example 2.5

Problem:

A well produces 270 gpm. If the drawdown for the well is 22 ft, what is the specific yield in gallons per minute per foot? What is the specific yield in gallons per minute per foot of drawdown?

Solution:

$$\text{Specific Yield, gpm/ft} = \frac{\text{Well Yield, gpm}}{\text{Drawdown, ft}}$$
$$= \frac{270 \text{ gpm}}{22 \text{ ft}}$$
$$= 12.3 \text{ gpm/ft}$$

Example 2.6

Problem:

The yield for a particular well is 300 gpm. If the drawdown for this well is 30 ft, what is the specific yield in gallons per minute per foot of drawdown?

Solution:

$$\text{Specific Yield, gpm/ft} = \frac{\text{Well Yield, gpm}}{\text{Drawdown, ft}}$$

$$= \frac{300 \text{ gpm}}{30 \text{ ft}}$$

$$= 10 \text{ gpm/ft}$$

2.2.4 WELL CASING DISINFECTION

A new, cleaned, or repaired well normally contains contamination which may remain for weeks unless the well is thoroughly disinfected. This may be accomplished by ordinary bleach in a concentration of 100 ppm (parts per million) of chlorine. The amount of disinfectant required is determined by the amount of water in the well. The following equation is used to calculate the pounds of chlorine required for disinfection:

$$\text{Chlorine, lbs} = (\text{Chlorine, mg/L}) \, (\text{Casing Vol., MG}) \, (8.34 \text{ lbs/gal}) \qquad (2.4)$$

Example 2.7

Problem:

A new well is to be disinfected with chlorine at a dosage of 50 mg/L. If the well casing diameter is 8 in and the length of the water-filled casing is 110 ft, how many pounds of chlorine will be required?

Solution:

First, calculate the volume of the water-filled casing:

$$(0.785) \, (0.67) \, (0.67) \, (110 \text{ ft}) \, (7.48 \text{ gal/cu ft}) = 290 \text{ gallons}$$

Then determine the pounds of chlorine required using the milligrams per liter to pounds equation:

$$\text{Chlorine, lbs} = (\text{chlorine, mg/L}) \text{ Volume, MG} \, (8.34 \text{ lbs/gal})$$

$$(50 \text{ mg/L}) \, (0.000290 \text{ MG}) \, (8.34 \text{ lbs/gal}) = 0.12 \text{ lbs Chlorine}$$

2.2.5 DEEP WELL TURBINE PUMP CALCULATIONS

The deep well turbine pump is used for high-capacity deep wells. The pump, consisting usually of more than one stage of centrifugal pump, is fastened to a pipe called the pump column; the pump is located in the water. The pump is driven from surface through a shaft running inside the pump column. The water is discharged from the pump up through the pump column to surface. The pump may be driven by a vertical shaft, by an electric motor at the top of the well, or by some other power source, usually through a right-angle gear drive located at the top of the well. A modern version of the deep well turbine pump is the submersible-type pump in which the pump, along with a close-coupled electric motor built as a single unit, is located below water level in the well. The motor is built to operate submerged in water.

2.2.6   VERTICAL TURBINE PUMP CALCULATIONS

The calculations pertaining to well pumps include head, horsepower, and efficiency calculations.

Discharge head is measured to the pressure gauge located close to the pump discharge flange. The pressure (in pounds per square inch) can be converted to feet of head using the equation:

$$\text{Discharge Head, ft} = (\text{press, psi}) \ (2.31 \text{ ft/psi}) \qquad (2.5)$$

Total pumping head (field head) is a measure of the lift below the discharge head pumping water level (discharge head). Total pumping head is calculated as follows:

$$\text{Pumping Head, ft} = \text{Pumping Water Level, ft} + \text{Discharge Head, ft} \qquad (2.6)$$

Example 2.8

Problem:

The pressure gauge reading at a pump discharge head is 4.2 psi. What is this discharge head expressed in feet?

Solution:

$$(4.2 \text{ psi}) \ (2.31 \text{ ft/psi}) = 9.7 \text{ ft}$$

Example 2.9

Problem:

The static water level of a pump is 100 ft. The well drawdown is 24 ft. If the gauge reading at the pump discharge head is 3.7 psi, what is the total pumping head?

Solution:

$$\begin{aligned}
\text{Total pumping head, ft} &= \text{Pumping water level, ft} + \text{discharge head, ft} \\
&= (100 \text{ ft} + 24 \text{ ft}) + (3.7 \text{ psi}) \ (2.31 \text{ ft/psi}) \\
&= 124 \text{ ft} + 8.5 \text{ ft} \\
&= 132.5 \text{ ft}
\end{aligned} \qquad (2.7)$$

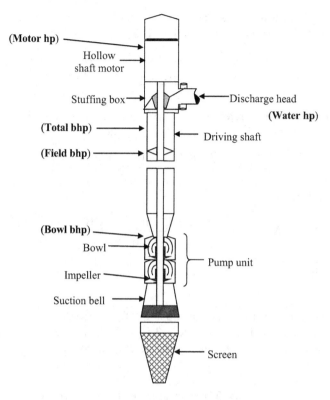

FIGURE 2.2 Vertical turbine pump, showing five horsepower types.

There are five types of horsepower calculations for vertical turbine pumps. It is important to have a general understanding of these five horsepower types (refer to Figure 2.2).

- **Motor horsepower** refers to the horsepower supplied to the motor. The following equation is used to calculate motor horsepower:

$$\text{Motor hp (input hp)} = \frac{\dfrac{\text{Field bhp}}{\text{Motor Efficiency}}}{100} \qquad (2.8)$$

- **Total brake horsepower** refers to horsepower output of the motor. The following equation is used to calculate total brake horsepower:

$$\text{Total bhp} = \text{Field bhp} + \text{Thrust Bearing Loss, hp} \qquad (2.9)$$

- **Field horsepower** refers to the horsepower required at the top of the pump shaft. The following equation is used to calculate field horsepower:

$$\text{Field bhp} = \text{Bowl bhp} + \text{Shaft Loss, hp} \qquad (2.10)$$

- **Bowl or laboratory horsepower** refers to the horsepower at the entry to the pump bowls. The following equation is used to calculate bowl horsepower:

$$\text{Bowl bhp (Lab bhp)} = \frac{(\text{Bowl Head, ft}) (\text{Capacity, gpm})}{\dfrac{(3960) (\text{Bowl Efficiency})}{100}} \qquad (2.11)$$

- **Water horsepower** refers to the horsepower at the pump discharge. The following equation is used to calculate water horsepower:

$$\text{Water hp} = \frac{(\text{Field Head, ft}) (\text{Capacity, gpm})}{3960} \qquad (2.12)$$

Or the equivalent equation:

$$\text{Water hp} = \frac{(\text{Field Head, ft}) (\text{Capacity, gpm})}{33,000 \text{ ft-lbs/min}}$$

Example 2.10

Problem:

The pumping water level for a well pump is 150 ft, and the discharge pressure measured at the pump discharge centerline is 3.5 psi. If the flow rate from the pump is 700 gpm, use equation 2.12 to calculate the water horsepower.

Solution:

First, calculate the field head. The discharge head must be converted from pounds per square inch to feet:

$$(3.5 \text{ psi}) (2.31 \text{ ft/psi}) = 8.1 \text{ ft}$$

The water horsepower is therefore:

$$150 \text{ ft} + 8.1 \text{ ft} = 158.1 \text{ ft}$$

The water horsepower can now be determined:

$$= \frac{(158.1 \text{ ft}) (700 \text{ gpm}) (8.34 \text{ lbs/gal})}{33,000 \text{ ft-lb/min}}$$

$$= 28 \text{ whp}$$

Example 2.11

Problem:

The pumping water level for a pump is 170 ft. The discharge pressure measured at the pump discharge head is 4.2 psi. If the pump flow rate is 800 gpm, use equation 2.12 to calculate the water horsepower.

Solution:

The field head must first be determined. In order to determine field head, the discharge head must be converted from pounds per square inch to feet:

$$(4.2 \text{ psi}) (2.31 \text{ ft/psi}) = 9.7 \text{ ft}$$

The field head can now be calculated:

$$170 \text{ ft} + 9.7 \text{ ft} = 179.7 \text{ ft}$$

And then the water horsepower can be calculated:

$$\text{Whp} = \frac{(179.7 \text{ ft}) (800 \text{ gpm})}{3960}$$
$$= 36 \text{ whp}$$

Example 2.12

Problem:

A deep well vertical turbine pump delivers 600 gpm. If the lab head is 185 ft and the bowl efficiency is 84%, use equation 2.11 to calculate the bowl horsepower.

$$\text{Bowl bhp} = \frac{(\text{Bowl Head, ft}) (\text{Capacity, gpm})}{(3960) \dfrac{(\text{Bowl Efficiency})}{100}}$$
$$= \frac{(185 \text{ ft}) (600 \text{ gpm})}{(3960) \dfrac{(84.0)}{100}}$$
$$= \frac{(185) (600 \text{ gpm})}{(3960) (0.84)}$$
$$= 33.4 \text{ bowl bhp}$$

Example 2.13

Problem:

The bowl brake horsepower is 51.8 bhp. If the 1 in diameter shaft is 170 ft long and is rotating at 960 rpm with a shaft fiction loss of 0.29 hp loss per 100 ft, what is the field brake horsepower?

Solution:

Before field brake horsepower can be calculated, the shaft loss must be factored in:

$$\frac{(0.29 \text{ hp loss})}{100} (170 \text{ ft}) = 0.5 \text{ hp loss}$$

Now the field brake horsepower can be determined:

$$\text{Field bhp} = \text{Bowl bhp} + \text{Shaft Loss, hp}$$
$$= 51.8 \text{ bhp} + 0.5 \text{ hp}$$
$$= 52.3 \text{ bhp}$$

Example 2.14

Problem:

The field horsepower for a deep well turbine pump is 62 bhp. If the thrust bearing loss is 0.5 hp and the motor efficiency is 88%, what is the motor input horsepower? (Use equation 15.8.)

Solution:

$$Mhp = \frac{\text{Total bhp}}{\frac{\text{Motor Efficiency}}{100}}$$

$$= \frac{62 \text{ bhp} + .0.5 \text{ hp}}{0.88}$$

$$= 71 \text{ mhp}$$

When we speak of the *efficiency* of any machine, we are speaking primarily of a comparison of what is put out by the machine (e.g., energy output) compared to its input (e.g., energy input). Horsepower efficiency, for example, is a comparison of horsepower output of the unit or system with horsepower input to that unit or system—the unit's efficiency. In regard to vertical turbine pumps, there are four types of efficiencies considered with vertical turbine pumps:

• Bowl efficiency
• Field efficiency
• Motor efficiency
• Overall efficiency

The general equation used in calculating percent efficiency is shown in the following.

$$\% = \frac{\text{Part}}{\text{Whole}} \times 100 \qquad (2.13)$$

Vertical turbine pump bowl efficiency is easily determined using a pump performance curve chart—provided by the pump manufacturer.

Field efficiency is determined by:

$$\text{Field Efficiency, } \% = \frac{\frac{(\text{Field head, ft}) (\text{Capacity, gpm})}{3960}}{\text{Total bhp}} \times 100 \qquad (2.14)$$

Example 2.15

Problem:
Given the following data, calculate the field efficiency of the deep well turbine pump.

Field head: 180 ft
Capacity: 850 gpm
Total brake horsepower: 61.3 bhp

Solution:

$$\text{Field Efficiency, } \% = \frac{(\text{Field head, ft}) (\text{Capacity, gpm})}{(3960) \text{Total bhp}} \times 100$$

$$= \frac{(180 \text{ ft}) (850 \text{ gpm})}{(3960) (61.3 \text{ bhp})} \times 100$$

$$= 63\%$$

The *overall efficiency* is a comparison of the horsepower output of the system with that entering the system. Equation 2.15 is used to calculate overall efficiency.

$$\text{Overall Efficiency, \%} = \frac{(\text{Field Efficiency, \%}) (\text{Motor Efficiency, \%})}{100} \qquad (2.15)$$

Example 2.16

Problem:

The efficiency of a motor is 90%. If the field efficiency is 83%, what is the overall efficiency of the unit?

Solution:

$$\text{Overall Efficiency, \%} = \frac{(\text{Field Efficiency, \%}) (\text{Motor Efficiency, \%})}{100} \times 100\%$$

$$= \frac{(83) (90)}{100}$$

$$= 74.7\%$$

2.3 WATER STORAGE

Water storage facilities for water distribution systems are required primarily to provide for fluctuating demands of water usage (to provide a sufficient amount of water to average or equalized daily demands on the water supply system). In addition, other functions of water storage facilities include increasing operating convenience, leveling pumping requirements (to keep pumps from running 24 hr a day), decreasing power costs, providing water during power source or pump failure, providing large quantities of water to meet fire demands, providing surge relief (to reduce the surge associated with stopping and starting pumps), increasing detention time (to provide chlorine contact time and satisfy the desired CT [contact time] value requirements), and blending water sources.

2.4 WATER STORAGE CALCULATIONS

The storage capacity, in gallons, of a reservoir, pond, or small lake can be estimated (see Figure 2.3) using equation 2.16.

$$\text{Reservoir, Pond or Lake Capacity, gal} = (\text{Ave. Length, ft})$$
$$(\text{Ave. Width, ft}) (\text{Ave. Depth, ft}) (7.48 \text{ gal/cu ft}) \qquad (2.16)$$

Example 2.17

Problem:

A pond has an average length of 250 ft, an average width of 110 ft, and an estimated average depth of 15 ft. What is the estimated volume of the pond in gallons?

(Top view of a pond)

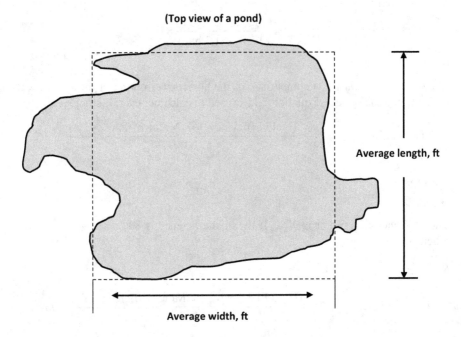

FIGURE 2.3   Determining pond storage capacity.

Solution:

$$\text{Vol. gal} = (\text{Ave. Length, ft}) (\text{Ave. Width, ft}) (\text{Ave. Depth, ft}) (7.48 \text{ gal/cu ft})$$
$$= (250 \text{ ft}) (110 \text{ ft}) (15 \text{ ft}) (7.48 \text{ gal/cu ft})$$
$$= 3,085,500 \text{ gal}$$

Example 2.18

Problem:

A small lake has an average length of 300 ft and an average width of 95 ft. If the maximum depth of the lake is 22 ft, what is the estimated gallons volume of the lake?

Note: For small ponds and lakes, the average depth is generally about 0.4 times the greatest depth. Therefore, to estimate the average depth, measure the greatest depth, then we multiply that number by 0.4.

Solution:

First, the average depth of the lake must be estimated:

$$\text{Estimated Aver. Depth, ft} = (\text{Greatest Depth, ft}) (0.4 \text{ Depth, ft})$$
$$= (22 \text{ ft}) (0.4 \text{ ft})$$
$$= 8.8 \text{ ft}$$

Then the lake volume can be determined:

Volume, gal = (Aver. Length, ft) (Aver. Width, ft) (Aver. Depth, ft) (7.48 gal/cu ft)

$$= (300 \text{ ft}) (95 \text{ ft}) (8.8 \text{ ft}) (7.48 \text{ cu ft})$$

$$= 1,875,984 \text{ gal}$$

2.5 COPPER SULFATE DOSING

Algal control is perhaps the most common in situ treatment of lakes, ponds, and reservoirs by application of copper sulfate—the copper ions in the water kill the algae. Copper sulfate application methods and dosages will vary depending on the specific surface water body being treated and also depends on alkalinity and pH. For example, if methyl orange alkalinity is <50 mg/L as $CaCO_3$, a dosage of 0.3 mg/L is recommended, based on total lake/reservoir volume. On the other hand, if methyl orange alkalinity is >50 mg/L as $CACO_3$, 1 mg/L for the upper 2 ft of the volume of the lake or reservoir is recommended. The desired copper sulfate dosage may be expressed in milligrams per liter copper, pounds copper sulfate per acre-foot, or pounds copper sulfate per acre. Potassium permanganate ($KMnO_4$) is added to ponds/reservoirs to oxidize iron and manganese and may be helpful in controlling algae.

For a dose expressed as milligrams per liter copper, the following equation is used to calculate pounds copper sulfate required:

$$\text{Copper Sulfate, lbs} = \frac{\dfrac{\text{Copper (mg/L) (Volume, MG) (8.34 lbs/gal)}}{\% \text{ Available Copper}}}{100} \quad (2.17)$$

Example 2.19

Problem:

For algae control in a small pond, a dosage of 0.5 mg/L copper is desired. The pond has a volume of 15 MG. How many pounds of copper sulfate will be required? (Copper sulfate contains 25% available copper.)

Solution:

$$\text{Copper Sulfate, lbs} = \frac{\dfrac{\text{(mg/L Copper) (Volume MG) (8.34 lbs/gal)}}{\% \text{ Available Copper}}}{100}$$

$$= \frac{\dfrac{(0.5 \text{ mg/L}) (15 \text{ MG}) (8.34 \text{ lbs/gal})}{25}}{100}$$

$$= 250 \text{ lbs Copper Sulfate}$$

For calculating pounds copper sulfate per acre-foot, use the following equation (assume the desired copper sulfate dosage is 0.9 lb/ac-ft):

$$\textit{Copper Sulfate, lbs} = \frac{(0.9 \text{ lbs Copper Sulfate) (ac-ft)}}{1 \text{ ac-ft}} \quad (2.18)$$

Example 2.20

Problem:

A pond has a volume of 35 ac-ft. If the desired copper sulfate dose is 0.9 lb/ac-ft, how many pounds of copper sulfate will be required?

Solution:

$$\text{Copper Sulfate, lbs} = \frac{(0.9 \text{ lb Copper Sulfate}) (\text{ac-ft})}{1 \text{ ac-ft}}$$

$$\frac{0.9 \text{ lbs Copper Sulfate}}{1 \text{ ac-ft}} = \frac{x \text{ lbs Copper Sulfate}}{35 \text{ ac-ft}}$$

Then solve for x:

$$(0.9)(35) = x$$

$$31.5 \text{ lb Copper Sulfate}$$

The desired copper sulfate dosage may also be expressed in terms of pounds copper sulfate per acre. The following equation is used to determine pounds copper sulfate (assume a desired dose of 5.2 lb copper sulfate/ac):

$$\text{Copper Sulfate, lbs} = \frac{(5.2 \text{ lb Copper Sulfate}) (\text{ac})}{1 \text{ ac}} \qquad (2.19)$$

Example 2.21

Problem:

A small lake has a surface area of 6 acres. If the desired copper sulfate dose is 5.2 lb/ac, how many pounds of copper sulfate are required?

Solution:

$$\text{Copper Sulfate, lb} = \frac{(5.2 \text{ lb copper sulfate}) (6.0 \text{ ac})}{1 \text{ ac}}$$

$$= 31.2 \text{ lb copper sulfate}$$

Example 2.22

Problem (a):

A holding pond measures 500 ft by 1,220 ft and has an average depth of 11 ft. What is the volume of the pond in acre-foot?

Solution (a):

$$\text{Area, ac} = \frac{\text{Length, ft} \times \text{Width, ft}}{43560 \text{ ft}^2/\text{ac}}$$

$$= \frac{(500 \text{ ft})(1220 \text{ ft})}{43560 \text{ ft}^2/\text{ac}} = 14.0 \text{ ac}$$

$$\text{Vol, ac-ft} = (14.0 \text{ ac})(11 \text{ ft}) = 154 \text{ ac-ft}$$

Problem (b):

What is the volume of the pond in million gallons?

Solution (b):

$$Vol, gal = (Vol, ac\text{-}ft)(43{,}560 \ ft^2/ac)(7.48 \ gal/ft^3)$$
$$= (154 \ ac\text{-}ft)(43{,}560 \ ft^2/ac)(7.48 \ gal/ft^3)$$
$$= 50{,}177{,}635 \ gal/1{,}000{,}000 \ gal/MG = 50.18 \ MG$$

Example 2.23

Problem:

For algae control of a reservoir, a dosage of 0.6 mg/L copper is desired. The reservoir has a volume of 21 MG. How many pounds of copper sulfate (25% available copper) will be required?

Solution:

$$lb = \frac{(dose)(volume)(8.34 \ lb/gal)}{\% \ copper}$$
$$= \frac{(0.6 \ mg/L)(21 \ MG)(8.34 \ lb/gal)}{0.25}$$
$$= 420.3 \ lb$$

Example 2.24

Problem:

The desired copper sulfate does in a reservoir is 4 mg/L. The reservoir has a volume of 60 ac-ft. How many pounds of copper sulfate (25% available copper) will be required?

Solution:

$$\frac{60 \ ac\text{-}ft}{} \ \frac{325828.8 \ gal}{1 \ ac\text{-}ft} \ \frac{1 \ MG}{1000000 \ gal} = 19.5 \ MG$$

$$lbs = \frac{(4 \ mg/L)(19.5 \ MG)(8.34)}{0.25}$$
$$= 2602.1 \ lbs$$

Example 2.25

Problem:

A pond has an average length of 260 ft, an average width of 80 ft, and an average depth of 12 ft. If the desired dose of copper sulfate is 0.9 lb/ac-ft, how many pounds of copper sulfate will be required?

Solution:

$$Vol, ac\text{-}ft = \frac{(260 \ ft)(80 \ ft)(12 \ ft)}{43560 \ ft^2/ac} = 5.7 \ ac\text{-}ft$$

$$lbs = \frac{5.7 \text{ ac-ft}}{} \frac{0.9 \text{ lb}}{\text{ac-ft}}$$

$$= 5.13 \text{ lb}$$

2.6 POTASSIUM PERMANGANATE DOSING

2.6.1 THE 411 ON IRON AND MANGANESE REMOVAL

Iron and manganese are frequently found in groundwater and in some surface waters. They do not cause health-related problems but are objectionable because they may cause aesthetic problems. Severe aesthetic problems may cause consumers to avoid an otherwise-safe water supply in favor of one of unknown or of questionable quality or may cause them to incur unnecessary expense for bottled water. Aesthetic problems associated with iron and manganese include the discoloration of water (iron = reddish water; manganese = brown or black water), staining of plumbing fixtures, imparting a bitter taste to the water, and stimulating the growth of microorganisms.

As mentioned, there is no direct health concerns associated with iron and manganese, although the growth of iron bacteria slimes may cause indirect health problems. Economic problems include damage to textiles, dye, paper, and food. Iron residue (or tuberculation) in pipes increases pumping head, decreases carrying capacity, may clog pipes, and may corrode through pipes.

Note: Iron and manganese are secondary contaminants. Their secondary maximum contaminant levels (SMCLs) are: iron = 0.3 mg/L; manganese = 0.05 mg/L.

Iron and manganese are most likely found in groundwater supplies, industrial waste, and acid mine drainage and as by-products of pipeline corrosion. They may accumulate in lake and reservoir sediments, causing possible problems during lake/reservoir turnover. They are not usually found in running waters (streams, rivers, etc.).

2.6.1.1 Iron and Manganese Removal Techniques

Chemical precipitation treatments for iron and manganese removal are called *deferrization* and *demanganization*. The usual process is *aeration*, where dissolved oxygen in the chemical causes precipitation; chlorine or potassium permanganate may be required.

2.6.1.2 Precipitation

Precipitation (or pH adjustment) of iron and manganese from water in their solid forms can be effected in treatment plants by adjusting the pH of the water by adding lime or other chemicals. Some of the precipitate will settle out with time, while the rest is easily removed by sand filters. This process requires pH of the water to be in the range of 10 to 11.

Note: Although the precipitation or pH adjustment technique for treating water containing iron and manganese is effective, note that the pH level must be adjusted higher (10–11 range) to cause the precipitation, which means that the pH level must then also be lowered (to the 8.5 range or a bit lower) to use the water for consumption.

2.6.1.3 Oxidation

One of the most common methods of removing iron and manganese is through the process of oxidation (another chemical process), usually followed by settling and filtration. Air, chlorine, or potassium permanganate can oxidize these minerals. Each oxidant has advantages and disadvantages, and each operates slightly differently:

1. *Air.* To be effective as an oxidant, the air must come in contact with as much of the water as possible. Aeration is often accomplished by bubbling diffused air through the water, by spraying the water up into the air, or by trickling the water over rocks, boards, or plastic packing materials in an aeration tower. The more finely divided the drops of water, the more oxygen comes in contact with the water and the dissolved iron and manganese.
2. *Chlorine.* This is one of the most popular oxidants for iron and manganese control because it is also widely used as a disinfectant; iron and manganese control by prechlorination can be as simple as adding a new chlorine feed point in a facility already feeding chlorine. It also provides a pre-disinfecting step that can help control bacterial growth through the rest of the treatment system. The downside to chorine use, however, is that when chlorine reacts with the organic materials found in surface water and some groundwaters, it forms TTHMs. This process also requires that the pH of the water be in the range of 6.5 to 7; because many groundwaters are more acidic than this, pH adjustment with lime, soda ash, or caustic soda may be necessary when oxidizing with chlorine.
3. *Potassium permanganate.* This is the best oxidizing chemical to use for manganese control removal. An extremely strong oxidant, it has the additional benefit of producing manganese dioxide during the oxidation reaction. Manganese dioxide acts as an adsorbent for soluble manganese ions. This attraction for soluble manganese provides removal to extremely low levels.

The oxidized compounds form precipitates that are removed by a filter. Note that sufficient time should be allowed from the addition of the oxidant to the filtration step. Otherwise, the oxidation process will be completed after filtration, creating insoluble iron and manganese precipitates in the distribution system.

2.6.1.4 Ion Exchange

The ion exchange process is used primarily to soften hard waters, but it will also remove soluble iron and manganese. The water passes through a bed of resin that adsorbs undesirable ions from the water, replacing them with less-troublesome ions. When the resin has given up all its donor ions, it is regenerated with strong salt brine (sodium chloride); the sodium ions from the brine replace the adsorbed ions and restore the ion exchange capabilities.

2.6.1.5 Sequestering

Sequestering or stabilization may be used when the water contains mainly low concentration of iron, and the volumes needed are relatively small. This process does not actually remove the iron or manganese from the water but complexes (binds it chemically) it with other ions in a soluble form that is not likely to come out of solution (i.e., not likely oxidized).

2.6.1.6 Aeration

The primary physical process uses air to oxidize the iron and manganese. The water is either pumped up into the air or allowed to fall over an aeration device. The air oxidizes the iron and manganese that is then removed by use of a filter. The addition of lime to raise the pH is often added to the process. While this is called a physical process, removal is accomplished by chemical oxidation.

2.6.1.7 Potassium Permanganate Oxidation and Manganese Greensand

The continuous regeneration potassium greensand filter process is another commonly used filtration technique for iron and manganese control. Manganese greensand is a mineral (gluconite) that has been treated with alternating solutions of manganous chloride and potassium permanganate. The result is a sand-like (zeolite) material coated with a layer of manganese dioxide—an adsorbent for soluble iron and manganese. Manganese greensand has the ability to capture (adsorb) soluble iron and manganese that may have escaped oxidation, as well as the capability of physically filtering out the particles of oxidized iron and manganese. Manganese greensand filters are generally set up as pressure filters, totally enclosed tanks containing the greensand. The process of adsorbing soluble iron and manganese "uses up" the greensand by converting the manganese dioxide coating to manganic oxide, which does not have the adsorption property. The greensand can be regenerated in much the same way as ion exchange resins, by washing the sand with potassium permanganate.

Example 2.26

Problem:

A chemical supplier recommends a 3% permanganate solution. If 2 lb $KMnO_4$ is dissolved in 12 gal of water, what is the percentage by weight?

Solution:

$$\frac{12 \text{ gal } 8.34 \text{ lb}}{\text{gal}} = 101 \text{ lb}$$

$$\% \text{ strength} = \frac{2.0 \text{ lb}}{101 \text{ lb} + 2.0 \text{ lb}} \times 100$$

$$= 1.94\%$$

Example 2.27

Problem:

To produce a 3% solution, how many pounds $KMnO_4$ should be dissolved in a tank 4 ft in diameter and filled to a depth of 4 ft?

$$\text{Vol, gal} = (0.785)(4.0)(4.0)(4.0)(7.48) = 375.8 \text{ gal}$$

$$\text{Chemical, lbs} = \frac{(\text{water vol})(8.34)(\% \text{ sol})}{(1 - \% \text{ sol})}$$

$$= \frac{(375.8 \text{ gal})(8.34)(0.03)}{1 - 0.03}$$

$$= 96.94 \text{ lb}$$

Example 2.28

Problem:

The plant's raw water has 1.6 mg/L of iron. How much $KMnO_4$ should be used to treat the iron? (Each 1 ppm of iron requires 0.91 mg/L of $KMnO_4$.)

Solution:

$$\frac{1.6 \text{ mg/L iron}}{} \frac{0.91 \text{ mg/L } KMnO_4}{1 \text{ ppm Fe}}$$

$$= 1.46 \text{ mg/L}$$

Example 2.29

Problem:

The plant's raw water has 6.8 mg/L of manganese. Hum much $KMnO_4$ should be used to treat manganese? (Each 1 ppm of manganese requires 1.92 mg/L of $KMnO_4$.)

Solution:

$$\text{mg/L } KMnO_4 = (6.8 \text{ mg/L})(1.92 \text{ mg/L})$$

$$= 13.06 \text{ mg/L}$$

Example 2.30

Problem:

The plant's raw water has 0.3 mg/L of iron and 2.6 mg/L of manganese. How much $KMnO_4$ should be used (0.91 mg/L $KMnO_4$ per 1.0 ppm Fe; 1.92 mg/L $KMnO_4$ per 1.0 ppm Mn)?

Solution:

$$\text{mg/L } KMnO_4 = (0.3 \text{ mg/L Fe})(0.91 \text{ mg/L}) = 0.273 \text{ mg/L}$$

$$\text{mg/L } KMnO_4 = 2.6 \text{ mg/L Mn})(1.92 \text{ mg/L}) = 4.99 \text{ mg/L}$$

$$0.273 + 4.99 = 5.26 \text{ mg/L}$$

Example 2.31

Problem:

A chemical supplier recommends a 3% permanganate solution mixed at a ratio of 0.22 lb/1 gal of water. How many milligrams $KMnO_4$ is there per milliliter of solution?

Solution:

$$\frac{0.22 \text{ lb}}{\text{gal}} \frac{1 \text{ gal}}{3785 \text{ mL}} \frac{453.69 \text{ g}}{1 \text{ lb}} \frac{1000 \text{ mg}}{1 \text{ g}} = 26.4 \text{ mg/mL}$$

Example 2.32

Problem:

A chemical supplier recommends a 3% permanganate solution mixed at a ratio of 0.25 lb/1 gal of water. If 60 gal of $KMnO_4$ are made at this ratio, how many pounds of chemical are required?

Solution:

$$lbs = \frac{0.25 \text{ lb}}{\text{gal}} \frac{60 \text{ gal}}{2}$$
$$= 15 \text{ lbs}$$

3 Coagulation and Flocculation Calculations

3.1 COAGULATION

Following screening and the other pretreatment processes, the next unit process in a conventional water treatment system is a mixer, where the first chemicals are added in what is known as coagulation. The exception to this situation occurs in small systems using groundwater, when chlorine or other taste and odor control measures are introduced at the intake and are the extent of treatment. The term *coagulation* refers to the series of chemical and mechanical operations by which coagulants are applied and made effective. These operations are comprised of two distinct phases: (1) rapid mixing to disperse coagulant chemicals by violent agitation into the water being treated and (2) flocculation to agglomerate small particles into well-defined floc by gentle agitation for a much longer time. The coagulant must be added to the raw water and perfectly distributed into the liquid; such uniformity of chemical treatment is reached through rapid agitation or mixing. Coagulation results from adding salts of iron or aluminum to the water and is a reaction between one of the following (coagulants) salts and water:

- Alum—aluminum sulfate
- Sodium aluminate
- Ferric sulfate
- Ferrous sulfate
- Ferric chloride
- Polymers

3.2 FLOCCULATION

Flocculation follows coagulation in the conventional water treatment process. *Flocculation* is the physical process of slowly mixing the coagulated water to increase the probability of particle collision. Through experience, we see that effective mixing reduces the required amount of chemicals and greatly improves the sedimentation process, which results in longer filter runs and higher-quality finished water. The goal of flocculation is to form a uniform, feather-like material similar to snowflakes—a dense, tenacious floc that entraps the fine, suspended, and colloidal particles and carries them down rapidly in the settling basin. To increase the speed of floc formation and the strength and weight of the floc, polymers are often added.

DOI: 10.1201/9781003354307-3

3.3 COAGULATION AND FLOCCULATION CALCULATIONS

In the proper operation of the coagulation and flocculation unit processes, calculations are performed to determine chamber or basin volume, chemical feed calibration, chemical feeder settings, and detention time.

3.3.1 CHAMBER AND BASIN VOLUME CALCULATIONS

To determine the volume of a square or rectangular chamber or basin, we use equation 3.1 or 3.2.

$$\text{Volume, cu ft} = (\text{length, ft}) (\text{width, ft}) (\text{depth, ft}) \qquad (3.1)$$

$$\text{Volume, gal} = (\text{length, ft}) (\text{width, ft}) (\text{depth, ft}) (7.48 \text{ gal/cu ft}) \qquad (3.2)$$

Example 3.1

Problem:

A flash mix chamber is 4 ft square with water to a depth of 3 ft. What is the volume of water (in gallons) in the chamber?

Solution:

$$\begin{aligned}
\text{Volume, gal} &= (\text{length, ft}) (\text{width, ft}) (\text{depth, ft}) (7.48 \text{ gal/cu ft}) \\
&= (4 \text{ ft}) (4 \text{ ft}) (3 \text{ ft}) (7.48 \text{ gal/cu ft}) \\
&= 359 \text{ gal}
\end{aligned}$$

Example 3.2

Problem:

A flocculation basin is 40 ft long, 12 ft wide, with water to a depth of 9 ft. What is the volume of water (in gallons) in the basin?

Solution:

$$\begin{aligned}
\text{Volume, gal} &= (\text{length, ft}) (\text{width, ft}) (\text{depth, ft}) (7.48 \text{ gal/cu ft}) \\
&= (40 \text{ ft}) (12 \text{ ft}) (9 \text{ ft}) (7.48 \text{ gal/cu ft}) \\
&= 32{,}314 \text{ gal}
\end{aligned}$$

Example 3.3

Problem:

A flocculation basin is 50 ft long, 22 ft wide, and contains water to a depth of 11 ft, 6 in. How many gallons of water are in the tank?

Solution:

First, convert the 6 in portion of the depth measurement to feet:

$$\frac{6 \text{ in.}}{12 \text{ in./ft}} = 0.5 \text{ ft}$$

Then use equation 3.2 to calculate basin volume:

$$\text{Volume, ft} = (\text{length, ft}) (\text{width, ft}) (\text{depth, ft}), (7.48 \text{ gal/cu ft})$$
$$= (50 \text{ ft}) (22 \text{ ft}) (11.5 \text{ ft}) (7.48 \text{ gal/cu ft})$$
$$= 94,622 \text{ gal}$$

3.3.2 DETENTION TIME

Because coagulation reactions are rapid, detention time for flash mixers is measured in seconds, whereas the detention time for flocculation basins is generally between 5 and 30 min. The equation used to calculate detention time is shown here.

$$\text{Detention Time, min} = \frac{\text{Volume of Tank, gal}}{\text{Flow Rate, gpm}} \qquad (3.3)$$

Example 3.4

Problem:

The flow to a flocculation basin 50 ft long, 12 ft wide, and 10 ft deep is 2,100 gpm. What is the detention time in the tank in minutes?

Solution:

$$\text{Tank Volume, gal} = (50 \text{ ft}) (12 \text{ ft}) (10 \text{ ft}) (7.48 \text{ gal/cu ft})$$
$$= 44,880 \text{ gal}$$

$$\text{Detention Time, min} = \frac{\text{Volume of Tank, gal}}{\text{Flow Rate, gpm}}$$
$$= \frac{44,880 \text{ gal}}{2100 \text{ gpm}}$$
$$= 21.4 \text{ min}$$

Example 3.5

Problem:

Assume the flow is steady and continuous for a flash mix chamber 6 ft long, 4 ft wide, with water to a depth of 3 ft. If the flow to the flash mix chamber is 6 MGD, what is the chamber detention time in seconds?

Solution:

First, convert the flow rate from gallons per day to gallons per second so that time units will match:

$$\frac{6,000,000}{(1440 \text{ min/day}) (60 \text{ sec/min})} = 69 \text{ gps}$$

Then calculate detention time using equation 3.3:

$$\text{Detention Time, sec} = \frac{\text{Volume of Tank, gal}}{\text{Flow Rate, gps}}$$

$$= \frac{(6 \text{ ft}) (4 \text{ ft}) (3 \text{ ft}) (7.48 \text{ gal/cu ft})}{69 \text{ gps}}$$

$$= 7.8 \text{ sec}$$

3.3.2.1 Determining Dry Chemical Feeder Setting (Lb/Day)

When adding (dosing) chemicals to the water flow, a measured amount of chemical is called for. The amount of chemical required depends on such factors as the type of chemical used, the reason for dosing, and the flow rate being treated. To convert from milligrams per liter to pounds per day, the following equation is used:

Chemical added, lbs/day = (Chemical, mg/L) (Flow, MGD) (8.34 lbs/gal)   (3.4)

Example 3.6

Problem:

Jar tests indicate that the best alum dose for a water is 8 mg/L. If the flow to be treated is 2,100,000 gpd, what should the pounds per day settling be on the dry alum feeder?

Solution:

$$\text{lbs/day} = (\text{Chemical, mg/L}) (\text{Flow, MGD}) (8.34 \text{ lbs/gal})$$

$$= (8 \text{ mg/L}) (2.10 \text{ MGD}) (8.34 \text{ lbs/gal})$$

$$= 140 \text{ lbs/day}$$

Example 3.7

Problem:

Determine the desired pounds per day setting on a dry chemical feeder if jar tests indicate an optimum polymer dose of 12 mg/L and the flow to be treated is 4.15 MGD.

Solution:

$$\text{Polymer, lbs/day} = (12 \text{ mg/L}) (4.15 \text{ MGD}) (8.34 \text{ lbs/gal})$$

$$= 415 \text{ lbs/day}$$

3.3.3   DETERMINING CHEMICAL SOLUTION FEEDER SETTING (GPD)

When solution concentration is expressed as pounds chemical per gallon solution, the required feed rate can be determined using the following equations:

$$\text{Chemical, lbs/day} = (\text{Chemical, mg/L}) \, (\text{Flow, MGD}) \, (8.34 \text{ lbs/gal}) \qquad (3.5)$$

Then convert the pounds per day dry chemical to gallons per day solution.

$$\text{Solution, gpd} = \frac{\text{Chemical, lbs/day}}{\text{lb Chemical/gal Solution}} \qquad (3.6)$$

Example 3.8

Problem:

Jar tests indicate that the best alum dose for a water is 7 mg/L. The water to be treated is 1.52 MGD. Determine the gallons per day setting for the alum solution feeder if the liquid alum contains 5.36 lb of alum per gallon of solution.

Solution:

First, calculate the pounds per day of dry alum required, using the milligrams per liter to pounds per day equation:

$$\text{Dry alum, lbs/day} = (\text{mg/L}) \, (\text{Flow, MGD}) \, (8.34 \text{ lb/gal})$$
$$= (7 \text{ mg/L}) \, (1.52 \text{ MGD}) \, (8.34 \text{ lb/gal})$$
$$= 89 \text{ lb/day}$$

Then, calculate gallons per day solution required.

$$\text{Alum Solution, gpd} = \frac{89 \text{ lb/day}}{5.36 \text{ lb alum/gal solution}}$$
$$= 16.6 \text{ gpd}$$

3.3.4   DETERMINING CHEMICAL SOLUTION FEEDER SETTING (ML/MIN)

Some solution chemical feeders dispense chemical as milliliters per minute (mL/min). To calculate the milliliter per minute solution required, use the following procedure:

$$\text{Solution, mL/min} = \frac{(\text{gpd}) \, (3785 \text{ mL/gal})}{1440 \text{ min/day}} \qquad (3.7)$$

Example 3.9

Problem:

The desired solution feed rate was calculated to be 9 gpd. What is this feed rate expressed as milliliters per minute?

Solution:

$$\text{mL/min} = \frac{(\text{gpd}) \, (3785 \text{ mL/gal})}{1440 \text{ min/day}}$$

$$= \frac{(9 \text{ gpd}) (3785 \text{ mL/gal})}{1440 \text{ min/day}}$$

$$= 24 \text{ mL/min Feed Rate}$$

Example 3.10

Problem:

The desired solution feed rate has been calculated to be 25 gpd. What is this feed rate expressed as milliliters per minute?

Solution:

$$\text{mL/min} = \frac{(\text{gpd}) (3785 \text{ mL/gal})}{1440 \text{ min/day}}$$

$$= \frac{(25 \text{ gpd}) (3785 \text{ mL/gal})}{1440 \text{ min/day}}$$

$$= 65.7 \text{ mL/min Feed Rate}$$

Sometimes we will need to know the milliliters per minute solution feed rate but we will not know the gallons per day solution feed rate. In such cases, calculate the gallons per day solution feed rate first, using the following the equation:

$$\text{Gpd} = \frac{(\text{Chemical, mg/L}) (\text{Flow, MGD}) (8.34 \text{ lb/gal})}{\text{Chemical, lb/Solution, gal}} \qquad (3.8)$$

3.3.5   DETERMINING PERCENT OF SOLUTIONS

The strength of a solution is a measure of the amount of chemical solute dissolved in the solution. We use the following equation to determine percent strength of solution using the following equation:

$$\% \text{ Strength} = \frac{\text{Chemical, lb}}{\text{Water, lb} + \text{Chemical, lb}} \times 100 \qquad (3.9)$$

Example 3.11

Problem:

If a total of 10 oz of dry polymer is added to 15 gal of water, what is the percent strength (by weight) of the polymer solution?

Solution:

Before calculating percent strength, the ounces chemical must be converted to pounds chemical:

$$\frac{10 \text{ ounces}}{16 \text{ ounces/pound}} = 0.625 \text{ lb chemical}$$

Now, calculate percent strength:

$$\% \text{ Strength} = \frac{\text{Chemical, lbs}}{\text{Water, lbs} + \text{Chemical, lbs}} \times 100$$

$$= \frac{0.625 \text{ lb chemical}}{(15 \text{ gal}) (8.34 \text{ lb/gal}) + 0.625 \text{ lb}} \times 100$$

$$= \frac{0.625 \text{ lb Chemical}}{125.7 \text{ lb Solution}} \times 100$$

$$= 0.5\%$$

Example 3.12

Problem:

If 90 g (1 g = 0.0022 lb) of dry polymer are dissolved in 6 gal of water, what percent strength is the solution?

Solution:

First, convert grams chemical to pounds chemical. Since 1 g equals 0.0022 lb, 90 g is 90 times 0.0022 lb:

$$(90 \text{ grams polymer}) (0.0022 \text{ lb/gram}) = 0.198 \text{ lb Polymer}$$

Now, calculate percent strength of the solution:

$$\% \text{ Strength} = \frac{\text{lbs Polymer}}{\text{lb Water} + \text{lb Polymer}} \times 100$$

$$= \frac{0.198 \text{ lb Polymer}}{(6 \text{ gal}) (8.34 \text{ lb/gal}) + 0.198 \text{ lb}} \times 100$$

$$= 4\%$$

3.3.6 DETERMINING PERCENT STRENGTH OF LIQUID SOLUTIONS

When using liquid chemicals to make up solutions (e.g., liquid polymer), a different calculation is required, as shown here:

$$\text{Liq. Poly., lb} \frac{\text{Liq. Poly (\% Strength)}}{100}$$

$$= \text{Poly. Sol., lb} \frac{\text{Poly. Sol. (\% Strength)}}{100} \tag{3.10}$$

Example 3.13

Problem:

A 12% liquid polymer is to be used in making up a polymer solution. How many pounds of liquid polymer should be mixed with water to produce 120 lb of a 0.5% polymer solution?

Solution:

$$(\text{Liq. Poly., lb}) \frac{(\text{Liq. Poly. \% Strength})}{100} = (\text{Poly Sol., lb}) \frac{(\text{Poly. Sol. \% Strength})}{100}$$

$$(x \text{ lbs}) \frac{(12)}{100} = (120 \text{ lbs}) \frac{(0.5)}{100}$$

$$x = \frac{(120)(0.005)}{0.12}$$

$$x = 5 \text{ lbs}$$

3.3.7 DETERMINING PERCENT STRENGTH OF MIXED SOLUTIONS

The percent strength of solution mixture is determined using the following equation:

$$\% \text{ Strength of Mix.} = \frac{(\text{Sol. 1, lbs})\dfrac{(\% \text{ Strength of Sol.1})}{100} + (\text{Sol. 2, lbs})\dfrac{(\% \text{ Strength of Sol 2})}{100}}{\text{lbs Solution 1} + \text{lbs Solution 2}} \times 100 \quad (3.11)$$

Example 3.14

Problem:

If 12 lb of a 10% strength solution are mixed with 40 lb of 1% strength solution, what is the percent strength of the solution mixture?

Solution:

$$\% \text{ Strength of Mix.} = \frac{(\text{Sol 1, lbs})\dfrac{(\% \text{ Strength, Sol 1})}{100} + (\text{Sol 2, lbs})\dfrac{(\% \text{ Strength, Sol 2})}{100}}{\text{lbs solution 1} + \text{lbs Solution 2}} \times 100$$

$$= \frac{(12 \text{ lb})(0.1) + (40 \text{ lb})(0.01)}{12 \text{ lb} + 40 \text{ lb}} \times 100$$

$$= \frac{1.2 \text{ lbs} + 0.40}{52 \text{ lb}} \times 100$$

$$= 3.1\%$$

3.3.8 DRY CHEMICAL FEEDER CALIBRATION

Occasionally we need to perform a calibration calculation to compare the actual chemical feed rate with the feed rate indicated by the instrumentation. To calculate the actual feed rate for a dry chemical feeder, place a container under the feeder, weigh the container when empty, then weigh the container again after a specified length of time (e.g. 30 min). The actual chemical feed rate can be calculated using the following equation:

$$\text{Chemical Feed Rate, lb/min} = \frac{\text{Chemical Applied, lb}}{\text{Length of Application, min}} \quad (3.12)$$

If desired, the chemical feed rate can be converted to pounds per day:

$$\text{Feed Rate, lb/day} = (\text{Feed Rate, lb/min}) \left(1440 \ \frac{\text{min}}{\text{day}}\right) \qquad (3.13)$$

Example 3.15

Problem:

Calculate the actual chemical feed rate, in pounds per day, if a container is placed under a chemical feeder and a total of 2 lb is collected during a 30 min period.

Solution:

First, calculate the pounds per minute feed rate:

$$\text{Chemical Feed Rate, lb/min} = \frac{\text{Chemical Applied, lb}}{\text{Length of Application, min}}$$

$$= \frac{2 \ \text{lb}}{30 \ \text{min}}$$

$$= 0.06 \ \text{lb/min Feed Rate}$$

Then calculate the pounds per day feed rate using equation 3.13:

$$\text{Chemical Feed Rate, lb/day} = (0.06 \ \text{lb/min}) (1440 \ \text{min/day})$$

$$= 86.4 \ \text{lb/day Feed Rate}$$

Example 3.16

Problem:

Calculate the actual chemical feed rate, in pounds per day, if a container is placed under a chemical feeder and a total of 1.6 lb is collected during a 20 min period.

Solution:

First, calculate the pounds per minute feed rate using equation 3.12:

$$\text{Chemical Feed Rate, lb/min} = \frac{\text{Chemical Applied, lb}}{\text{Length of Application, min}}$$

$$= \frac{1.6 \ \text{lb}}{20 \ \text{min}}$$

$$= 0.08 \ \text{lb/min Feed Rate}$$

Then calculate the pounds per day feed rate:

$$\text{Chemical Feed Rate, lb/day} = (0.08 \ \text{lb/min}) (1440 \ \text{min/day})$$

$$= 115 \ \text{lb/day Feed Rate}$$

3.3.9 SOLUTION CHEMICAL FEEDER CALIBRATION

As with other calibration calculations, the actual solution chemical feed rate is determined and then compared with the feed rate indicated by the instrumentation. To calculate the actual solution chemical feed rate, first express the solution feed rate in million gallons per day. Once the million gallons per day solution flow rate has been calculated, use the milligrams per liter equation to determine chemical dosage in pounds per day.

If solution fed is expressed as milliliters per minute, first convert milliliters per minute flow rate to gallons per day flow rate.

$$gpd = \frac{(mL/min)(1440 \; min/day)}{3785 \; mL/gal} \tag{3.14}$$

Then calculate chemical dosage, in pounds per day.

$$\text{Chemical, lb/day} = (mg/L \; Chemical) \; (MGD \; Flow) \; (8.34 \; lb/day) \tag{3.15}$$

Example 3.17

Problem:

A calibration test is conducted for a solution chemical feeder. During a 5 min test, the pump delivered 940 mg/L of the 1.20% polymer solution. (Assume the polymer solution weighs 8.34 lb/gal.) What is the polymer dosage rate in pounds per day?

Solution:

The flow rate must be expressed as million gallons per day. Therefore, the milliliters per minute solution flow rate must first be converted to gallons per day and then million gallons per day. The milliliters per minute flow rate is calculated as:

$$\frac{940 \; mL}{5 \; min} = 188 \; mL/min$$

Next, convert the milliliters per minute flow rate to gallons per day flow rate:

$$\frac{(188 \; mL/min) \; (1440 \; min/day)}{3785 \; mL/gal} = 72 \; gpd \; flow \; rate$$

Then calculate the pounds per day polymer feed rate:

$$(12,000 \; mg/L) \; (0.000072 \; MGD) \; (8.34 \; lb/day) = 7.2 \; lb/day \; Polymer$$

Example 3.18

Problem:

A calibration test is conducted for a solution chemical feeder. During a 24 hr period, the solution feeder delivers a total of 100 gal of solution. The polymer solution is a 1.2% solution. What is the pounds per day feed rate? (Assume the polymer solution weighs 8.34 lb/gal.)

Solution:

The solution feed rate is 100 gpd (or 100 gallons per day). Expressed as million gallons per day, this is 0.000100 MGD. Use equation 3.15 to calculate actual feed rate in pounds per day:

$$\text{lbs/day Chemical} = (\text{Chemical, mg/L)}\,(\text{Flow, MGD)}\,(8.34\text{ lb/day})$$
$$= (12,000\text{ mg/L})\,(0.000100\text{ MGD})\,(8.34\text{ lb/day})$$
$$= 10\text{ lb/day Polymer}$$

The actual pumping rates can be determined by calculating the volume pumped during a specified time frame. For example, if 60 gal are pumped during a 10 min test, the average pumping rate during the test is 6 gpm.

Actual volume pumped is indicated by drop in tank level. By using the following equation, we can determine the flow rate in gallons per minute.

$$\text{Flow Rate, gpm} = \frac{(0.785)\,(D^2)\,(\text{Drop in Level, ft})\,(7.48\text{ gal/cu ft})}{\text{Duration of Test, min}} \qquad (3.16)$$

Example 3.19

Problem:

A pumping rate calibration test is conducted for a 15 min period. The liquid level in the 4 ft diameter solution tank is measured before and after the test. If the level drops 0.5 ft during the 15 min test, what is the pumping rate in gallons per minute?

$$\text{Flow Rate, gpm} = \frac{(0.785)\,(D^2)\,(\text{Drop, ft})\,(7.48\text{ gal/cu ft})}{\text{Duration of Test, min}}$$
$$= \frac{(0.785)\,(4\text{ ft})\,4\text{ ft})\,(0.5\text{ ft})\,(7.48\text{ gal/cu ft})}{15\text{ min}}$$
$$= 3.1\text{ gpm Pumping Rate}$$

3.3.10 DETERMINING CHEMICAL USAGE

One of the primary functions performed by water operators is the recording of data. The pounds per day or gallons per day chemical use is part of this data. From this data, the average daily use of chemicals and solutions can be determined. This information is important in forecasting expected chemical use, comparing it with chemicals in inventory, and determining when additional chemicals will be required. To determine average chemical use, we use equation 3.17 (in pounds per day) or equation 3.18 (in gallons per day).

$$\text{Average Use, lb/day} = \frac{\text{Total Chemical Used, lb}}{\text{Number of days}} \qquad (3.17)$$

or

$$\text{Average Use, gpd} = \frac{\text{Total Chemical Used, gal}}{\text{Number of Days}} \qquad (3.18)$$

Then we can calculate days' supply in inventory:

$$\text{Days' Supply in Inventory} = \frac{\text{Total Chemical in Inventory, lb}}{\text{Average Use, lb/day}} \qquad (3.19)$$

or

$$\text{Days' Supply in Inventory} = \frac{\text{Total Chemical in Inventory, gal}}{\text{Average Use, gpd}} \qquad (3.20)$$

Example 3.20

Problem:

The chemical used for each day during a week is given in the following. Based on this data, what was the average pounds per day chemical use during the week?

Monday: 88 lb/day   Friday: 96 lb/day
Tuesday: 93 lb/day   Saturday: 92 lb/day
Wednesday: 91 lb/day   Sunday: 86 lb/day
Thursday: 88 lb/day

Solution:

$$\text{Average Use, lb/day} = \frac{\text{Total Chemical Used, lb}}{\text{Number of Days}}$$

$$= \frac{634 \text{ lb}}{7 \text{ days}}$$

$$= 90.6 \text{ lb/day Average Use}$$

Example 3.21

Problem:

The average chemical use at a plant is 77 lb/day. If the chemical inventory is 2,800 lb, how many days' supply is this?

Solution:

$$\text{Days' Supply in Inventory} = \frac{\text{Total Chemical in Inventory, lb}}{\text{Average Use, lb/day}}$$

$$= \frac{2800 \text{ lb in Inventory}}{77 \text{ lb/day Average Use}}$$

$$= 36.4 \text{ days' Supply in Inventory}$$

Note: Jar tests are performed as required on settling tank influent and are beneficial in determining the best flocculant aid and appropriate doses to improve solids capture during periods of poor settling.

3.4 COAGULATION AND FLOCCULATION PRACTICE PROBLEMS

Problem 3.1

The average flow for a water plant is 5 MGD. A jar test indicates that the best alum dosage is 3.2 mg/L. How many pounds per day will the operator feed?

Solution:

$$lb/day = (dose)(flow)(8.34 \text{ lb/gal})$$
$$= 3.2 \text{ mg/L})(5.0 \text{ MGD})(8.34 \text{ lb/gal})$$
$$= 133.4 \text{ lb/day}$$

Problem 3.2

If the average flow for a waterworks is 1,100,000 gal per day. A jar test indicates that the best alum dosage is 1.6 mg/L. How many grams per minute should the feeder deliver?

Solution:

$$gram/min = \frac{(dose)(flow)(3.785 \text{ L/gal})}{(1440 \text{ min/day})(1000 \text{ mg/g})}$$
$$= \frac{(1.6 \text{ mg/L})(1100000 \text{ gpd})(3.785)}{(1440)(1000)} = 4.63 \text{ g/mi}$$

Problem 3.3

A waterworks used 40 lb of alum treating 3.5 MGD. Calculate the dose in milligrams per liter.

Solution:

$$dose, mg/L = \frac{\text{feed rate, lb/day}}{(flow)(8.34 \text{ lb/gal})}$$
$$= \frac{40 \text{ lb}}{(3.5 \text{ MGD})(8.34 \text{ lb/gal})}$$
$$= 1.37 \text{ mg/L}$$

Problem 3.4

Liquid polymer is supplied to a water treatment plant as an 8% solution. How many gallons of this liquid is required to make 140 gal of 1.5% polymer solution?

Solution:

$$C_1 V_1 = C_2 V_2$$
$$(0.08)(V1) = (0.015)(140 \text{ gal})$$
$$V1 = \frac{(0.015)(140 \text{ gal})}{0.08}$$
$$= 26.25 \text{ gal}$$

Problem 3.5

Liquid alum delivered to a waterworks contains 801.5 mg of alum per milliliter of liquid solution. Jar tests indicate that the best alum dose is 5 mg/L. Determine the setting on the liquid alum feeder in milliliters per minute if that flow is 2.5 MGD.

Solution:

$$ml/min = \frac{(dose)(flow, gpd)(3.785 \text{ L/gal})}{(conc, mg/mL)(1440 \text{ min/day})}$$

$$= \frac{(5 \text{ mg/L})(2500000 \text{ gpd})(3.785)}{(801.5 \text{ mg/mL})(1440)} = 41 \text{ mL/min}$$

Problem 3.6

A waterworks operator switches from dry alum to liquid alum. If she feeds an average of 160 lb of dry alum a day, how many gallons of liquid alum will she need to feed on average given the following information?
Given:

Alum liquid: 48% concentration
10.5 lb dry alum/gallon (weight)
5.5 lb dry alum/gallon (concentration)
1.335 specific gravity

Solution:

$$\frac{160 \text{ lb}}{} \quad \frac{gal}{5.5 \text{ lb alum}}$$
$$= 29.1 \text{ gal}$$

Problem 3.7

The average flow for a waterworks is 3.5 MGD. A jar test indicates that the best alum dosage is 2.2 mg/L. How man pounds per day will the operator feed?

Solution:

$$lb/day = (dose)(flow)(8.34 \text{ lb/gal})$$
$$= (2.2 \text{ mg/L})(3.5 \text{ MGD})(8.34 \text{ lb/gal})$$
$$= 64.22 \text{ lb/day}$$

Problem 3.8

The average flow for a water treatment plant is 12.8 MGD. The jar test indicates the best alum dose is 1.8 mg/L. How many pounds per day will the operator feed?

Solution:

$$lb/day = (1.8 \text{ mg/L})(12.8 \text{ MGD})(8.34 \text{ lb/gal})$$
$$= 192.2 \text{ lb/day}$$

Problem 3.9

Determine the setting on a dry alum feeder in pounds per day when the flow is 1.4 MGD. Jar tests indicate that the best alum dose is 14 mg/L.

Solution:

$$lb/day = (14 \text{ mg/L})(1.4 \text{ MGD})(8.34 \text{ lb/gal})$$
$$163.5 \text{ lb/day}$$

Problem 3.10

The average flow for a water plant is 8.5 MGD. A jar test indicates that the best alum dosage is 2.4 mg/L. How many grams per minute should the feeder deliver?

Solution:

$$gram/min = \frac{(\text{dose})(\text{flow})(3.785 \text{ L/gal})}{(1440 \text{ min/day})(1000 \text{ mg/g})}$$
$$= \frac{(2.4 \text{ mg/L})(8500000 \text{ gpd})(3.785 \text{ L/gal})}{(1440 \text{ m/d})(1000 \text{ mg/g})}$$
$$= 53.6 \text{ gram/min}$$

Problem 3.11

The average daily flow for a waterworks is 0.85 MGD. If the polymer dosage is kept at 1.9 mg/L, how many pounds of polymer will be used in 30 days?

Solution:

$$lb/day = (1.9 \text{ mg/L})(0.85)(8.34 \text{ lb/gal})$$
$$= (13.47 \text{ lb/day})(30 \text{ days})$$
$$= 404.1 \text{ lb}$$

Problem 3.12

The average flow for a water plant is 9,050 gpm. A jar test indicates that the best polymer dose is 3.2 mg/L. How many pounds will the plant feed in one week? (Assume the plant runs 24 hr/day, 7 days/week.)

Solution:

$$\frac{9,050 \text{ gpm}}{\text{min}} \frac{1440 \text{ min}}{\text{day}} \frac{1 \text{ MG}}{1000000} = 13.03 \text{ MGD}$$

$$lb/day = (3.2 \text{ m/L})(13.03 \text{ MGD})(8.34)$$
$$= (347.74 \text{ lb/day})(7 \text{ days/wk})$$
$$= 2434.18 \text{ lb/week}$$

Problem 3.13

A water treatment plant used 30 lb of cationic polymer to treat 1.8 MG of water during a 24 hr period. What is the polymer dosage in milligrams per liter?

Solution:

$$mg/L = \frac{\text{feed rate}}{(\text{flow})(8.34)}$$

$$= \frac{30 \text{ lb/day}}{(1.8 \text{ MGD})(8.34 \text{ lb/gal})}$$

$$= 2.0 \text{ mg/L (rounded)}$$

Problem 3.14

A waterworks fed 140 lb of alum is treating 1.5 MGD. Calculate the dose in milligrams per liter.

Solution:

$$mg/L = \frac{130 \text{ lb/day}}{(1.5 \text{ MGD}) (8.34 \text{ lb/gal})}$$

$$= 10.4 \text{ mg/L}$$

Problem 3.15

A waterworks fed 50 g/min of dry alum while treating 2.8 MGD. Calculate the milligrams per liter dose.

Solution:

$$\frac{2.8 \text{ MGD}}{\text{day}} \quad \frac{\text{day}}{1440 \text{ min}} \quad \frac{1000000 \text{ gal}}{1 \text{ MG}} = 1944.4 \text{ gpm}$$

$$mg/L = \frac{(\text{feed rate, g/min})(1000 \text{ mg/g})}{(\text{flow, gpm})(3,785 \text{ L/gal})}$$

$$= \frac{(50 \text{ grams})(1000 \text{ mg/g})}{(1944.4 \text{ gpm})(3.785 \text{ L/gal})}$$

$$= 6.8 \text{ mg/L}$$

Problem 3.16

Liquid polymer is supplied to a waterworks as an 8% solution. How many gallons of this liquid polymer should be used to make 220 gal of a 0.8% polymer solution?

Solution:

$$C_1 V_1 = C_2 V_2$$

$$(0.08)(V_1) = (0.008)(220 \text{ gal})$$

$$V_1 = \frac{(0.008)(220 \text{ gal})}{0.08}$$

$$= 22 \text{ gal}$$

Problem 3.17

Liquid polymer is supplied to a water treatment plant as a 7% solution. How many gallons of this liquid polymer should be used to make 6 gal of a 6% polymer solution?

Solution:

$$C_1V_1 = C_2V_2$$
$$(0.07)(V_1) = (0.06)(6\ \text{gal})$$
$$V_1 = \frac{(0.06)(6\ \text{gal})}{0.07}$$
$$= 5.14\ \text{gal}$$

Problem 3.18

Liquid polymer is supplied to a water treatment plant as a 7% solution. How many gallons of liquid polymer should be used to make 60 gal of a 0.5% polymer solution?

Solution:

$$(0.07)(V_1) = (0.005)(60\ \text{gal})$$
$$V_1 = \frac{(0.005)(60\ \text{gal})}{0.07}$$
$$V_1 = 4.29\ \text{gal}$$

Problem 3.19

Liquid alum delivered to a waterworks contains 653.4 mg of alum per milliliter of liquid solution. Jar tests indicate that the best alum dose is 9 mg/L. Determine the setting on the liquid alum chemical feed in milliliters per minute if the flow is 2.3 MGD.

Solution:

$$\text{mL/min} = \frac{(\text{dose mg/L})(\text{flow, gpd})(3.785\ \text{L/gal})}{(\text{conc, mg/mL})(1440\ \text{min/day})}$$
$$= \frac{(9\ \text{mg/L})(2300000)(3.785\ \text{L/gal})}{(653.4\ \text{mg/mL})(1440\ \text{min/day})}$$
$$= 83.3\ \text{mL/min}$$

Problem 3.20

Three 2 min samples are collected from an alum dry feeder. What is the feed rate in milligrams per liter when the flow rate is 3 MGD?

Given:

Sample 1 = 22 g
Sample 2 = 25 g
Sample 3 = 23 g

Solution:

$$\frac{22g + 25g + 23g}{3} = 23.3 \text{ g}$$

$$\frac{3 \text{ MG}}{\text{Day}} \frac{\text{day}}{1440 \text{ min}} \frac{1000000 \text{ gal}}{1 \text{ MG}} = 2083.3 \text{ gpm}$$

$$mg/L = \frac{(\text{feed rate})(1000 \text{ mg/g})}{(\text{flow})(3.785 \text{ L/gal})}$$

$$= \frac{(23.3g/2min)(1000mg/g)}{(2083.3 \text{ gpm})(3.783L/gal)}$$

$$= 2.95 \text{ mg/L}$$

Problem 3.21

A waterworks is treating 9.1 MGD with 2.0 mg/L liquid alum. How many gallons per day of liquid alum will be required? The liquid alum contains 5.25 lb dry alum per gallon.

$$gal/day = \frac{(\text{dose})(\text{flow})(8.34 \text{ lb/gal})}{\text{conc. lb/gal}}$$

$$= \frac{(2.0 \text{ mg/L})(9.1 \text{ MGD})(8.34 \text{ lb/gal})}{5.25 \text{ lb/gal}}$$

$$= 28.91 \text{ gal/day}$$

Problem 3.22

A jar test indicates the 3.2 mg/L of liquid alum is required in treating 7.5 MGD. How many milliliters per minute should the metering pump deliver? The liquid alum delivered to the plant contains 640 mg alum per milliliters of liquid solution.

Solution:

$$mL/min = \frac{(3.2 \text{ mg/L})(7500000 \text{ gal/day})(3.785 \text{ L/gal})}{(640 \text{ mg/mL})(1440 \text{ min/day})}$$

$$= 98.57 \text{ mL/min}$$

Problem 3.23

A jar test indicates that 1.9 mg/L of liquid ferric chloride should be fed to treat 2,878 gpm of water. How many milliliters per minute should be fed by a metering pump? Ferric chloride contains 4.6 lb dry chemical per gallon of liquid solution.

Solution:

$$\frac{2878 \text{ gal}}{\text{min}} \frac{1440 \text{ min}}{\text{day}} = 4144320 \text{ gpd}$$

$$\frac{4.6 \text{ lb}}{\text{gal}} \frac{1 \text{ gal}}{3785 \text{ mL}} \frac{453.6\text{g}}{1 \text{ lb}} \frac{1000 \text{ mg}}{1 \text{ g}} = 551.27 \text{ mg/mL}$$

$$\text{mL/min} = \frac{(1.9 \text{ mg/L})(4144320)(3.785)}{(551.27 \text{ mg/mL})(1440 \text{ min/d})} = 37.54 \text{ mL/min}$$

3.5   CHEMICAL FEEDERS

Simply put, a *chemical feeder* is a mechanical device for measuring a quantity of chemical and applying it to water at a preset rate.

Types of Chemical Feeders

Two types of chemical feeders are commonly used: solution (or liquid) feeders and dry feeders. Liquid feeders apply chemicals in solutions or suspensions, and dry feeders apply chemicals in granular or powdered forms. In a solution feeder, chemical enters the feeder and leaves the feeder in a liquid state; in a dry feeder, chemical enters and leaves the feeder in a dry state.

Solution Feeders

Solution feeders are small positive displacement metering pumps of three types: (1) reciprocating (piston plunger or diaphragm types), (2) vacuum type (e.g., gas chlorinator), or (3) gravity feed rotameter (e.g., drip feeder). Positive displacement pumps are used in high-pressure, low-flow applications; they deliver a specific volume of liquid for each stroke of a piston or rotation of an impeller.

Dry Feeders

Two types of dry feeders are volumetric and gravimetric, depending on whether the chemical is measured by volume (volumetric type) or weight (gravimetric type). Simpler and less-expensive than gravimetric pumps, volumetric dry feeders are also less-accurate. Gravimetric dry feeders are extremely accurate, deliver high feed rates, and are more expensive than volumetric feeders.

Chemical Feeder Calibration

Chemical feeder calibration ensures effective control of the treatment process. Obviously, chemical feed without some type of metering and accounting of chemical used adversely affects the water treatment process. Chemical feeder calibration also optimizes economy of operation; it ensures the optimum use of expensive chemicals. Finally, operators must have accurate knowledge of each individual feeder's capabilities at specific settings. When a certain dose must be administered, the operator must rely on the feeder to feed the correct amount of chemical. Proper calibration ensures chemical dosages can be set with confidence. At a minimum, chemical feeders must be calibrated on an annual basis. During operation, when the operator changes chemical strength or chemical purity or makes any adjustment to the feeder, or when the treated water flow changes, the chemical feeder should be calibrated. Ideally, anytime maintenance is performed on chemical feed equipment, calibration should be performed.

What factors affect chemical feeder calibration (i.e., feed rate)? For solution feeders, calibration is affected anytime solution strength changes, anytime a mechanical change is introduced in the pump (change in stroke length or stroke frequency), and/or whenever flow rate changes. In the dry chemical feeder, calibration is affected anytime chemical purity changes, mechanical damage occurs (e.g., belt change), and/or whenever flow rate changes. In the calibration process, calibration charts are usually used or made up to fit the calibration equipment. The calibration chart is also affected by certain factors, including change in chemical, change in flow rate of water being treated, and/or a mechanical change in the feeder.

Note: Pounds per day (lb/day) is not normally useful information for setting the feed rate setting on a feeder. This is the case because process control usually determines a dosage in parts per million, milligrams per liter, or grains per gallon. A separate chart may be necessary for another conversion based on the individual treatment facility flow rate.

3.6   CHEMICAL FEEDER CALCULATIONS

Example 3.22

Problem:

An operator collects three 3 min samples from a dry feeder. Based on the information given, determine the average grams per minute.

Given:

Sample 1 = 36.8 g
Sample 2 = 37.8 g
Sample 3 = 35.9 g

$$\text{Average} = \frac{36.8 \text{ g} + 37.8 \text{ g} + 35.9 \text{ g}}{3} = 36.8 \text{ g}$$

Solution:

$$\text{gram/min} = \frac{36.8 \text{ g}}{3} = 12.28 \text{ g/min}$$

Example 3.23

Problem:

What is the average dose in milligrams per liter for the feeder in the previous question if the plant treats 3.7 MGD?

Solution:

$$\frac{3700000 \text{ gal}}{\text{day}} \frac{\text{day}}{60 \text{ min}} = 61{,}667 \text{ gpm}$$

$$\text{dose, mg/L} = \frac{(\text{feedrate, g/min})(1000 \text{ mg/g})}{(\text{flow, gpm})(3.785 \text{ L/gal})}$$

$$= \frac{(12.28 \text{ g/min})(1000 \text{ mg/g})}{(61{,}667 \text{ gpm})(3.785 \text{ L/gal})} = 0.053$$

Example 3.24

Problem:

An operator is checking the calibration on a chemical feeder. The feeder delivers 105 g in 5 min. How many grams per minutes does the feeder deliver? How many pounds per day does the feeder deliver?

Solution:

$$\text{Gram/min} = \frac{105 \text{ grams}}{5 \text{ min}}$$

$$= 21 \text{ g/min}$$

$$\frac{21 \text{ g}}{\text{min}} \frac{1440 \text{ min}}{\text{day}} \frac{1 \text{ lb}}{453.6 \text{ g}} = 66.7 \text{ lb/day}$$

Example 3.25

Problem:

An operator checks the calibration of a dry feeder by catching samples and weighing them on a balance. Each catch lasts 1 min. Calculate the average feed rate in grams per minute based on the following data, and determine how many pounds per hour is being fed?

Given:

Sample 1 weighs 38.0 g.
Sample 2 weighs 36.3 g.
Sample 3 weighs 39.2 g.
Sample 4 weighs 38.5 g.

Solution:

$$\text{Avg} = \frac{38.0 \text{ g} + 36.3 \text{ g} + 39.2 \text{ g} + 38.5 \text{ g}}{4} = 38 \text{ g}$$

$$\text{Gram/min} = 38 \text{ g}/1 \text{ min} = 38/\text{min}$$

$$\frac{38 \text{ g}}{\text{min}} \frac{60 \text{ min}}{\text{hr}} \frac{1 \text{ lb}}{453.6 \text{ g}} = 5.03 \text{ lb/hr}$$

Example 3.26

Problem:

An operator collects three 2 min samples form a dry feeder:

Sample 1 weighs 21.3 g.
Sample 2 weighs 24.2 g.
Sample 3 weighs 21.7 g.

What is the average grams per minute? What is the average dose in milligrams per liter for the feet if the plant treats 430,000 gpd?

Solution:

$$\text{Avg} \frac{21.3\ \text{g} + 24.2\ \text{g} + 21.7\ \text{g}}{3} = 22.4\ \text{g}$$

$$\text{gram/min} \frac{22.4\ \text{g}}{2\ \text{min}} = 11.2\ \text{g/min}$$

$$\frac{430000\ \text{gal}}{\text{day}} \frac{1\ \text{day}}{1440\ \text{min}} = 298.6\ \text{gpm}$$

$$\text{mg/L} = \frac{(11.2\ \text{g/min})(1000\ \text{mg/g})}{(298.6\ \text{gpm})(3.785\ \text{L/gal})} = 9.9\ \text{mg/L}$$

Example 3.27

Problem:

An operator collects five 2 min samples form a dry feeder:

Sample 1 weighs 49.4 g.
Sample 2 weighs 44.2 g.
Sample 3 weighs 41.8 g.
Sample 4 weighs 48.4 g.
Sample 5 weighs 47.9 g.

What is the average grams per minute? What is the average does in milligrams per liter if the plant treats 1,300,000 gpd?

Solution:

$$\text{Avg} = \frac{49.4\ \text{g} + 44.2\ \text{g} + 41.8\ \text{g} + 48.4\ \text{g} + 47.9\ \text{g}}{5} = 46.34\ \text{g}$$

$$\text{gram/min} = \frac{46.34\ \text{g}}{2\ \text{min}} = 23.17\text{/min}$$

$$\frac{1300000\ \text{gal}}{\text{Day}} \frac{1\ \text{day}}{1440\ \text{min}} = 902.8\ \text{gpm}$$

$$\text{mg/L} = \frac{(23.17\text{g/min})(1000\ \text{mg/g})}{(902.8\ \text{gpm})(3.785\ \text{L/gal})} = 6.78\ \text{mg/l}$$

Example 3.28

Problem:

A chemical feeder calibration is tested using a 1,000 mL graduated cylinder. The cylinder is filled to 830 mL in a 3 min test. What is the chemical feed rate in milliliters per minute? What is the chemical feed rate in gallons per minute? What is the chemical feed rate in gallons per day?

Solution:

$$\text{mL/min} = \frac{830 \text{ mL}}{3} = 276.7 \text{ mL/min}$$

$$\frac{276.7 \text{ mL}}{\text{min}} \frac{\text{L}}{1000 \text{ mL}} \frac{1 \text{ gal}}{3.785 \text{ L}} = 0.071 \text{ gal/min}$$

$$\frac{0.071 \text{ gal}}{\text{min}} \frac{1440 \text{ min}}{1 \text{ day}} = 102.24 \text{ gpd}$$

4 Sedimentation Calculations

4.1 SEDIMENTATION

Sedimentation, the solid–liquid separation by gravity, is one of the most basic processes of water and wastewater treatment. In water treatment, plain sedimentation, such as the use of a presedimentation basin for grit removal and sedimentation basin following coagulation–flocculation, is the most commonly used.

4.2 TANK VOLUME CALCULATIONS

The two common shapes of sedimentation tanks are rectangular and cylindrical. The equations for calculating the volume for each type of tank are shown in the following.

4.2.1 CALCULATING TANK VOLUME

For rectangular sedimentation basins, we use equation 4.1.

$$\text{Volume, gal} = (\text{length, ft}) \ (\text{width, ft}) \ (\text{depth, ft}) \ (7.48 \text{ gal/cu ft}) \qquad (4.1)$$

For circular clarifiers, we use equation 4.2.

$$\text{Volume, gal} = (0.785) \ (D^2) \ (\text{depth, ft}) \ (7.48 \text{ gal/cu ft}) \qquad (4.2)$$

Example 4.1

Problem:

A sedimentation basin is 25 ft wide, 80 ft long, and contains water to a depth of 14 ft. What is the volume of water in the basin in gallons?

Solution:

$$\text{Volume, gal} = (\text{length, ft}) \ (\text{width, ft}) \ (\text{depth, ft}) \ (7.48 \text{ gal/cal ft})$$
$$= (80 \text{ ft}) \ (25 \text{ ft}) \ (14 \text{ ft}) \ (7.48 \text{ gal/cu ft})$$
$$= 209{,}440 \text{ gal}$$

Example 4.2

Problem:

A sedimentation basin is 24 ft wide and 75 ft long. When the basin contains 140,000 gal, what would the water depth be?

DOI: 10.1201/9781003354307-4

Solution:

$$\text{Volume, gal} = (\text{length, ft}) \ (\text{width, ft}) \ (\text{depth, ft}) \ (7.48 \text{ gal/cu ft})$$

$$140,000 \text{ gal} = (75 \text{ ft}) \ (24 \text{ ft}) \ (x \text{ ft}) \ (7.48 \text{ gal/cu ft})$$

$$x \text{ ft} = \frac{140,000}{(75)(24)(7.48)}$$

$$x \text{ ft} = 10.4 \text{ ft}$$

4.3 DETENTION TIME (DT)

Detention time for clarifiers varies from 1 to 3 hr. The equations used to calculate detention time are shown here.

Basic detention time equation:

$$\text{Detention Time, hrs} = \frac{\text{Volume of Tank, gal}}{\text{Flow Rate, gph}} \qquad (4.3)$$

Rectangular sedimentation basin equation:

$$\text{Detention Time, hrs} = \frac{(\text{Length, ft}) \ (\text{Width, ft}) \ (\text{Depth, ft}) \ (7.48 \text{ gal/cu ft})}{\text{Flow Rate, gph}} \qquad (4.4)$$

Circular basin equation:

$$\text{Detention Time, hr} = \frac{(0.785) \ (D^2) \ (\text{Depth, ft}) \ (7.48 \text{ gal/cu ft})}{\text{Flow Rate, gph}} \qquad (4.5)$$

Example 4.3

Problem:

A sedimentation tank has a volume of 137,000 gal. If the flow to the tank is 121,000 gph, what is the detention time in the tank in hours?

Solution:

$$\text{Detention Time, hrs} = \frac{\text{Volume of Tank, gal}}{\text{Flow Rate, gph}}$$

$$= \frac{137,000 \text{ gal}}{121,000 \text{ gph}}$$

$$= 1.1 \text{ hours}$$

Example 4.4

Problem:

A sedimentation basin is 60 ft long, 22 ft wide, and has water to a depth of 10 ft. If the flow to the basin is 1,500,000 gpd, what is the sedimentation basin detention time?

Solution:

First, convert the flow rate from gallons per day to gallons per hour so that time units will match (1,500,000 gpd ÷ 24 hrs/day = 62,500 gph). Then calculate detention time:

Volume of tank, in galllons
Detention time, in hours = ----------------------------
Flow rate, in gallons per hour
(60 ft) (22 ft) (10 ft) (7.48 gal/cu ft)
= ---
62,500 gph
= 1.6 hr

$$\text{Detention Time, hrs} = \frac{\text{Volume of Tank, gal}}{\text{Flow Rate, gph}}$$

$$= \frac{(60\text{ ft})\ (22\text{ ft})\ (10\text{ ft})\ (7.48\text{ gal/cu ft})}{62,500\text{ gph}}$$

$$= 1.6\text{ hours}$$

4.4 SURFACE OVERFLOW RATE (SOR)

Surface loading rate—similar to hydraulic loading rate (flow per unit area)—is used to determine loading on sedimentation basins and circular clarifiers. Hydraulic loading rate, however, measures the total water entering the process, whereas surface overflow rate measures only the water overflowing the process (plant flow only).

- **Note:** Surface overflow rate calculations do not include recirculated flows. Other terms used synonymously with *surface overflow rate* are *surface loading rate* and *surface settling rate*.

Surface overflow rate is determined using the following equation:

$$\text{Surface Overflow Rate} = \frac{\text{Flow, gpm}}{\text{Area, sq ft}} \qquad (4.6)$$

Example 4.5

Problem:

A circular clarifier has a diameter of 80 ft. If the flow to the clarifier is 1,800 gpm, what is the surface overflow rate in gallons per minute per square foot?

Solution:

$$\text{Surface Overflow Rate} = \frac{\text{Flow, gpm}}{\text{Area, sq ft}}$$

$$= \frac{1800\text{ gpm}}{(0.785)\ (80\text{ ft})\ (80\text{ ft})}$$

$$= 0.36\text{ gpm/sq ft}$$

Example 4.6

Problem:

A sedimentation basin 70 ft by 25 ft receives a flow of 1,000 gpm. What is the surface overflow rate in gallons per minute per square foot?

Solution:

$$\text{Surface Overflow Rate} = \frac{\text{Flow, gpm}}{\text{Area, sq ft}}$$

$$= \frac{1,000 \text{ gpm}}{(70 \text{ ft}) \ (25 \text{ ft})}$$

$$= 0.6 \text{ gpm/sq ft}$$

4.5 MEAN FLOW VELOCITY

The measure of average velocity of the water as it travels through a rectangular sedimentation basin is known as **mean flow velocity**. Mean flow velocity is calculated using equation 4.7.

$$Q \ (\text{Flow}), \text{cu ft/min} = A \ (\text{Cross-Sectional Area}), \text{ft}^2 \times V \ (\text{Volume}) \ \text{ft/min} \quad (4.7)$$

$$(Q = A \times V)$$

Example 4.7

Problem:

A sedimentation basin 60 ft long and 18 ft wide has water to a depth of 12 ft. When the flow through the basin is 900,000 gpd, what is the mean flow velocity in the basin in feet per minute?

Solution:

Because velocity is desired in feet per minute, the flow rate in the Q = AV equation must be expressed in cubic feet per minute (cfm):

$$\frac{900,000 \text{ gpd}}{(1440 \text{ min/day}) \ (7.48 \text{ gal/cu ft})} = 84 \text{ cfm}$$

Then use the Q = AV equation to calculate velocity:

$$Q = AV$$

$$84 \text{ cfm} = (18 \text{ ft}) \ (12 \text{ ft}) \ (\text{xfpm})$$

$$x = \frac{84}{(18) \ (12)}$$

$$= 0.4 \text{ fpm}$$

Example 4.8

Problem:

A rectangular sedimentation basin 50 ft long and 20 ft wide has a water depth of 9 ft. If the flow to the basin is 1,880,000 gpd, what is the mean flow velocity in feet per minute?

Solution:

Because velocity is desired in feet per minute, the flow rate in the Q = AV equation must be expressed in cubic feet per minute (cfm):

$$\frac{1,880,000 \text{ gpd}}{(1440 \text{ min/day}) \ (7.48 \text{ gal/cu ft})} = 175 \text{ cfm}$$

Then use the Q = AV equation to calculate velocity:

$$Q = AV$$
$$175 \text{ cfm} = (20 \text{ ft}) \ (9 \text{ ft}) \ (x \text{ fpm})$$
$$x = \frac{175}{(20) \ (9)}$$
$$x = 0.97 \text{ fpm}$$

4.6   WEIR LOADING RATE (WEIR OVERFLOW RATE, WOR)

Weir loading rate (weir overflow rate) is the amount of water leaving the settling tank per linear foot of weir. The result of this calculation can be compared with design. Normally, weir overflow rates of 10,000 to 20,000 gal/day/ft are used in the design of a settling tank. Typically, weir loading rate is a measure of the gallons per minute (gpm) flow over each foot (ft) of weir. Weir loading rate is determined using the following equation:

$$\text{Weir Loading Rate, gpm/ft} = \frac{\text{Flow, gpm}}{\text{Weir Length, ft}} \qquad (4.8)$$

Example 4.9

Problem:

A rectangular sedimentation basin has a total of 115 ft of weir. What is the weir loading rate in gallons per minute per foot when the flow of 1,110,000 gpd?

Solution:

$$\frac{1,110,000 \text{ gpd}}{1440 \text{ min/day}} = 771 \text{ gpm}$$

$$\text{Weir Loading Rate} = \frac{\text{Flow, gpm}}{\text{Weir Length, ft}}$$

$$= \frac{771 \text{ gpm}}{115 \text{ ft}}$$

$$= 6.7 \text{ gpm/ft}$$

Example 4.10

Problem:

A circular clarifier receives a flow of 3.55 MGD. If the diameter of the weir is 90 ft, what is the weir loading rate in gallons per minute per foot?

Solution:

$$\frac{3,550,000 \text{ gpd}}{1440 \text{ min/day}} = 2465 \text{ gpm}$$

$$\text{ft of weir} = (3.14) \ (90 \text{ ft})$$

$$= 283 \text{ ft}$$

$$\text{Weir Loading Rate} = \frac{\text{Flow, gpm}}{\text{Weir Length, ft}}$$

$$= \frac{2465 \text{ gpm}}{283 \text{ ft}}$$

$$= 8.7 \text{ gpm/ft}$$

4.7   PERCENT SETTLED BIOSOLIDS

The percent settled biosolids test (a.k.a. "volume over volume" test, or V/V test) is conducted by collecting a 100 mL slurry sample from the solids contact unit and allowing it to settle for 10 min. After 10 min, the volume of settled biosolids at the bottom of the 100 mL graduated cylinder is measured and recorded. The equation used to calculate percent settled biosolids is shown in the following.

$$\% \text{ Settled Biosolids} = \frac{\text{Settled Biosolids Volume, Ml}}{\text{Total Sample Volume, Ml}} \times 100 \qquad (4.9)$$

Example 4.11

Problem:

A 100 mL sample of slurry from a solids contact unit is placed in a graduated cylinder and allowed to set for 10 min. The settled biosolids at the bottom of the graduated cylinder after 10 min is 22 mL. What is the percent of settled biosolids of the sample?

Solution:

$$\% \text{ Settled Biosolids} = \frac{\text{Settled Biosolids, Ml}}{\text{Total Sample, Ml}} \times 100$$

$$= \frac{22 \text{ Ml}}{100 \text{ Ml}} \times 100$$

$$= 19\% \text{ Settled Biosolids}$$

Example 4.12

Problem:

A 100 mL sample of slurry from a solids contact unit is placed in a graduated cylinder. After 10 min, a total of 21 mL of biosolids settled to the bottom of the cylinder. What is the percent settled biosolids of the sample?

$$\% \text{ Settled Biosolids} = \frac{\text{Settled Biosolids, Ml}}{\text{Total Sample, Ml}} \times 100$$

$$= \frac{21 \text{ Ml}}{100 \text{ Ml}} \times 100$$

$$= 21\% \text{ Settled Biosolids}$$

4.8   DETERMINING LIME DOSAGE (MG/L)

During the alum dosage process, lime is sometimes added to provide adequate alkalinity (HCO_3) in the solids contact clarification process for the coagulation and precipitation of the solids. To determine the lime dose required, in milligrams per liter, three steps are required.

In *step 1*, the total alkalinity required is calculated. Total alkalinity required to react with the alum to be added and provide proper precipitation is determined using the following equation:

Total Alk. Required, mg/L = Alk. Reacting with Alum, mg/L + Alk.

In the Water, mg/L   (4.10)

↑

$$\left(1 \text{ mg/L alum reacts w/0.45 mg/L Alk.}\right)$$

Example 4.13

Problem:

A raw water requires an alum dose of 45 mg/L, as determined by jar testing. If a residual 30 mg/L alkalinity must be present in the water to ensure complete precipitation of alum added, what is the total alkalinity required in milligrams per liter?

Solution:

First, calculate the alkalinity that will react with 45 mg/L alum:

$$\frac{0.45 \text{ mg/L Alk}}{1 \text{ mg/L Alum}} = \frac{X \text{ mg/L Alk}}{45 \text{ mg/L Alum}}$$

$$(0.45)\ (45) = x$$
$$= 20.25 \text{ mg/L Alk.}$$

Next, calculate the total alkalinity required:

Total Alk. Required, mg/L = Alk to React w/Alum, mg/L + Residual Alk, mg/L
$$= 20.25 \text{ mg/L} + 30 \text{ mg/L}$$
$$= 50.25 \text{ mg/L}$$

Example 4.14

Problem:

Jar tests indicate that 36 mg/L alum are optimum for particular raw water. If a residual 30 mg/L alkalinity must be present to promote complete precipitation of the alum added, what is the total alkalinity required in milligrams per liter?

Solution:

First, calculate the alkalinity that will react with 36 mg/L alum:

$$\frac{0.45 \text{ mg/L Alk}}{1 \text{ mg/L Alum}} = \frac{X \text{ mg/L Alk}}{36 \text{ mg/L Alum}}$$
$$(0.45)\ (36) = x$$
$$= 16.2$$

Then, calculate the total alkalinity required:

$$\text{Total Alk. Required, mg/L} = 16.2 \text{ mg/L} + 30 \text{ mg/L}$$
$$= 46.2 \text{ mg/L}$$

In **step 2**, we make a comparison between required alkalinity and alkalinity already in the raw water to determine how many milligrams per liter alkalinity should be added to the water. The equation used to make this calculation is shown here:

Alk. To be Added to the Water, mg/L = Tot. Alk. Req'd, mg/L – Alk.
$$\text{Present in the Water, mg/L} \qquad (4.11)$$

Example 4.15

Problem:

A total of 44 mg/L alkalinity is required to react with alum and ensure proper precipitation. If the raw water has an alkalinity of 30 mg/L as bicarbonate, how many milligrams per liter alkalinity should be added to the water?

Solution:

Alk. To be added, mg/L = Total Alk. Req'd, mg/L – Alk.
$$\text{Present in the Water, mg/L}$$
$$= 44 \text{ mg/L} - 30 \text{ mg/L}$$
$$= 14 \text{ mg/L Alkalinity to be Added}$$

In *step 3*, after determining the amount of alkalinity to be added to the water, we determine how much lime (the source of alkalinity) needs to be added. We accomplish this by using the ratio shown in example 4.16.

Example 4.16

Problem:

It has been calculated that 16 mg/L alkalinity must be added to a raw water. How much milligrams per liter lime will be required to provide this amount of alkalinity? (1 mg/L alum reacts with 0.45 mg/L, and 1 mg/L alum reacts with 0.35 mg/L lime.)

Solution:

First, determine the milligrams per liter lime required by using a proportion that relates bicarbonate alkalinity to lime:

$$\frac{0.45 \; mg / L \; Alk}{0.35 \; mg/L \; Lime} = \frac{16 \; mg / L \; Alk}{x \; mg/L \; Lime}$$

Next, we cross-multiply:

$$0.45x = (16)\,(0.35)$$
$$x = \frac{(16)\,(0.35)}{0.45}$$
$$x = 12.4 \; mg/L \; Lime$$

In example 4.17, we use all three steps to determine lime dosage (in milligrams per liter) required.

Example 4.17

Problem:

Given the following data, calculate the lime dose required, in milligrams per liter:

Alum dose required (determined by jar tests): 52 mg/L.
Residual alkalinity required for precipitation: 30 mg/L.
1 mg/L alum reacts with 0.35 mg/L lime.
1 mg/L alum reacts with 0.45 mg/L alkalinity.
Raw water alkalinity: 36 mg/L.

Solution:

To calculate the total alkalinity required, you must first calculate the alkalinity that will react with 52 mg/L alum:

$$\frac{0.45 \; mg/L \; Alk}{1 \; mg/L \; Alum} = \frac{X mg/L \; Alk}{52 \; mg/L \; Alum}$$
$$(0.45)\,(52) = x$$
$$23.4 \; mg/L \; Alk = x$$

The total alkalinity requirement can now be determined:

Total Alk. Required, mg/L = Alk. To React w/Alum, mg/L + Residual Alk, mg/L

$$= 23.4 \text{ mg/L} + 30 \text{ mg/L}$$

$$= 53.4 \text{ mg/L Total Alkalinity Required}$$

Next, calculate how much alkalinity must be **added** to the water:

Alk. To be Added, mg/L = Total Alk Required, mg/L – Alk. Present, mg/L

$$= 53.4 \text{ mg/L} - 36 \text{ mg/L}$$

$$= 17.4 \text{ mg/L Alk to be added to the Water}$$

Finally, calculate the lime required to provide this additional alkalinity:

$$\frac{0.45 \text{ mg/L Alk}}{0.35 \text{ mg/L Lime}} = \frac{17.4 \text{ mg/L Alk}}{x \text{ mg/L Lime}}$$

$$0.45x = (17.4)\ (0.35)$$

$$x = \frac{(17.4)(0.35)}{0.45}$$

$$x = 13.5 \text{ mg/L Lime}$$

4.9   DETERMINING LIME DOSAGE (LB/DAY)

After the lime dose has been determined in terms of milligrams per liter, it is a fairly simple matter to calculate the lime dose in pounds per day, which is one of the most common calculations in water and wastewater treatment. To convert from milligrams per liter to pounds per day lime dose, we use the following equation:

$$\text{Lime, lb/day} = \text{Lime } (\text{mg/L})\ (\text{Flow, MGD})\ (8.34 \text{ lb/gal}) \qquad (4.12)$$

Example 4.18

Problem:

The lime dose for a raw water has been calculated to be 15.2 mg/L. If the flow to be treated is 2.4 MGD, how many pounds per day lime will be required?

Solution:

$$\text{Lime, lbs/day} = (\text{Lime, mg/L})\ (\text{Flow, MGD})\ (8.34 \text{ lbs/gal})$$

$$= (15.2 \text{ mg/L})\ (2.4 \text{ MGD})\ (8.34 \text{ lbs/gal})$$

$$= 304 \text{ lbs/day Lime}$$

Example 4.19

Problem:

The flow to a solids contact clarifier is 2,650,000 gpd. If the lime dose required is determined to be 12.6 mg/L, how many pounds per day lime will be required?

Solution:

$$\begin{aligned} \text{Lime, lbs/day} &= \left(\text{Lime, mg/L}\right) \ \left(\text{Flow, MGD}\right) \ \left(8.34 \text{ lbs/gal}\right) \\ &= \left(12.6 \text{ mg/L}\right) \ \left(2.65 \text{ MGD}\right) \ \left(8.34 \text{ lbs/gal}\right) \\ &= 278 \text{ lbs/day Lime} \end{aligned}$$

4.10   DETERMINING LIME DOSAGE (G/MIN)

In converting from milligrams per liter lime to grams per minute (g/min) lime, use equation 4.13.

Key Point: 1 lb = 453.6 g.

$$\text{Lime, g/min} = \frac{\left(\text{Lime, lb/day}\right) \ \left(453.6 \text{ g/lb}\right)}{1440 \text{ min/day}} \tag{4.13}$$

Example 4.20

Problem:

A total of 275 lb/day lime will be required to raise the alkalinity of the water passing through a solids-contact clarification process. How many grams per minute lime does this represent?

Solution:

$$\begin{aligned} \text{Lime, g/min} &= \frac{\left(\text{lb/day}\right) \ \left(453.6 \text{ g/lb}\right)}{1440 \text{ min/day}} \\ &= \frac{\left(275 \text{ lb/day}\right) \ \left(453.6 \text{ g/lb}\right)}{1440 \text{ min/day}} \\ &= 86.6 \text{ g/min Lime} \end{aligned}$$

Example 4.21

Problem:

A lime dose of 150 lb/day is required for a solids-contact clarification process. How many grams per minute lime does this represent?

Solution:

$$\begin{aligned} \text{Lime, g/min} &= \frac{\left(\text{lb/day}\right) \ \left(453.6 \text{ g/lb}\right)}{1440 \text{ min/day}} \\ &= \frac{\left(150 \text{ lb/day}\right) \ \left(453.6 \text{ g/lb}\right)}{1440 \text{ min/day}} \\ &= 47.3 \text{ g/min Lime} \end{aligned}$$

4.11   SEDIMENTATION PRACTICE PROBLEMS

Problem 4.1

The flow to a sedimentation tanks is 220,000 gpd. If the tank is 52 ft long and 34 ft wide, what is the surface overflow rate in gallons per day per square foot?

Solution:

$$SOR = \frac{\text{flow, gpd}}{\text{area, ft}^2}$$

$$SOR = \frac{220000 \text{ gpd}}{(52 \text{ ft})(34 \text{ ft})} = 124.4 \text{ gpd/ft}^2$$

Problem 4.2

A tank has a length of 80 ft and is 22 ft wide. What is the weir length around the basin in feet?

Solution:

$$\text{Weir length} = (2)(\text{length}) + (2)(\text{width})$$
$$= (2)(80 \text{ ft}) + (2)(22 \text{ ft})$$
$$= 160 \text{ ft} + 44 \text{ ft}$$
$$= 204 \text{ ft}$$

Problem 4.3

A clarifier has a diameter of 80 ft. What is the length of the weir around the clarifier in feet?

Solution:

$$\text{Weir length} = (3.14)(\text{diameter})$$
$$= (3.14)(80 \text{ ft})$$
$$= 251.2 \text{ ft}$$

Problem 4.4

The diameter of weir in a circular clarifier is 100 ft. What is the weir overflow rate in gallons per day per foot if the flow over the weir is 1.80 MGD?

Solution:

$$WOR = \frac{\text{flow, gpd}}{\text{weir length, ft}}$$
$$= \frac{1,800,000 \text{ gpd}}{(31.4)(100 \text{ ft})}$$
$$= 5732.5 \text{ gpd/ft}$$

Problem 4.5

A clarifier is 42 ft long, 30 ft wide, and 12 ft deep. If the daily flow is 3.3 MGD, what is the detention time (in minutes) in the basin?

Solution:

$$\text{Volume} = (42 \text{ ft})(30 \text{ ft})(12 \text{ ft})(7.48) = 113{,}098 \text{ gal}$$

$$\text{DT, hr} = \frac{(\text{vol, gal})(24 \text{ hr/day})}{\text{flow, gpd}}$$

$$= \frac{(113098 \text{ gal})(24 \text{ hr/day})}{3{,}300{,}000 \text{ gpd}}$$

$$= (0.823 \text{ hr})(60 \text{ min/hr}) = 49.4 \text{ min}$$

Problem 4.6

A tank has a length of 110 ft, a width of 22 ft, and a depth of 12 ft. What is the surface area in square feet?

Solution:

$$A = (\text{L, ft})(\text{W, ft})$$

$$A = (110 \text{ ft})(22 \text{ ft})$$

$$A = 2420 \text{ ft}^2$$

Problem 4.7

A clarifier has a diameter of 80 ft and a depth of 10 ft. What is the surface area of the clarifier in square feet?

Solution:

$$A = (0.785)(\text{d, ft})^2$$

$$A = (0.785)(80 \text{ ft})(80 \text{ ft})$$

$$A = 5024 \text{ ft}^3$$

Problem 4.8

The flow to a sedimentation tank is 3.15 MGD. If the tank is 80 ft long and 22 ft wide, what is the surface overflow rate in gallons per day per square foot?

Solution:

$$\text{SOR} = \frac{\text{flow, gal/day}}{\text{area, ft}^2}$$

$$= \frac{3{,}150{,}000 \text{ gpd}}{(80 \text{ ft})(22 \text{ ft})}$$

$$= 1789.8 \text{ gpd/ft}^2$$

Problem 4.9

The flow to a sedimentation tank is 55,000 gpd. If the tank is 50 ft long and 14 ft wide, what is the surface overflow rate (in gallons per day per square foot)?

Solution:

$$\text{SOR} = \frac{55,000 \text{ gpd}}{(50 \text{ ft})(14 \text{ ft})}$$

$$= 79 \text{ gpd/ft}^2$$

Problem 4.10

A sedimentation tank is 88 ft long and 42 ft wide and receives a flow of 5.05 MGD. Calculate the SOR in gallons per day per square foot.

Solution:

$$\text{SOR} = \frac{5050000 \text{ gpd}}{(88 \text{ ft})(42 \text{ ft})}$$

$$= 1366.3 \text{ gpd/ft}^2$$

Problem 4.11

A circular clarifier has a diameter of 70 ft. If the flow to the clarifier is 3.5 MGD, what is the surface overflow rate (in gallons per day per square foot)?

Solution:

$$\text{SOR} = \frac{3500000 \text{ gpd}}{(0.785)(70 \text{ ft})(70 \text{ ft})}$$

$$= 909.92 \text{ gpd/ft}^2$$

Problem 4.12

A clarifier has a flow rate of 4,800 gpm and a diameter of 80 ft. What is the surface overflow rate in gallons per day per square foot?

Solution:

$$\frac{4800 \text{ gal}}{\text{min}} \frac{1440 \text{ min}}{\text{day}} = 6912000 \text{ gpd}$$

$$\text{SOR} = \frac{6912000 \text{ gpd}}{(0.785)(80 \text{ ft})(80 \text{ ft})}$$

$$= 1375.8 \text{ gpd/ft}^2$$

Problem 4.13

A clarifier with a diameter of 52 ft receives a flow of 2.085 MGD. What is the surface overflow rate (in gallons per day per square foot)?

Solution:

$$SOR = \frac{2085000 \text{ gpd}}{(0.785)(52 \text{ ft})(52 \text{ ft})}$$

$$= 982.3 \text{ gpd/ft}^2$$

Problem 4.14

What is the gallons per day per square foot overflow to a circular clarifier that has the following:

Diameter: 80 ft
Flow: 1,960 gpm

$$\frac{1960 \text{ gal}}{\text{min}} \frac{1440 \text{ min}}{\text{day}} = 2822400 \text{ gpd}$$

$$SOR = \frac{2822400 \text{ gpd}}{(0.785)(80 \text{ ft})(80 \text{ ft})} = 562 \text{ gpd/ft}^2$$

Problem 4.15

A rectangular clarifier receives a flow of 5.5 MGD. The length of the clarifier is 96 ft, 6 in, and the width is 76 ft, 6 in. What is the SOR in gallons per day per square foot?

Solution:

$$SOR = \frac{5500000 \text{ gpd}}{(96.5 \text{ ft})(76.5 \text{ ft})} = 744.95 \text{ gpd/ft}^2$$

Problem 4.16

A tank has a length of 110 ft, a width of 26 ft, and a depth of 12 ft. What is the weir length around the basin in feet?

Solution:

$$\text{Weir length} = 2(\text{Weir length}) + 2(\text{Weir width})$$

$$= 2(110 \text{ ft}) + 2(26 \text{ ft}) = 220 \text{ ft} + 52 \text{ ft}$$

$$= 272 \text{ ft}$$

Problem 4.17

A clarifier has a diameter of 80 ft and a depth of 14 ft. What is the length of the weir around the clarifier in feet?

Solution:

$$\text{Weir length} = (3.14)(\text{Weir diameter})$$

$$= (3.14)(80 \text{ ft})$$

$$= 251.2 \text{ ft}$$

Problem 4.18

A sedimentation tank has a total of 140 ft of weir over which the water flows. What is the weir overflow rate in gallons per day per foot of weir when the flow is 1.8 MGD?

Solution:

$$WOR = \frac{flow, gpd}{weir\ length, ft}$$

$$= \frac{1,800,000\ gpd}{140\ ft}$$

$$= 12,857.14\ gpd/ft$$

Problem 4.19

The diameter of the weir in a circular clarifier is 90 ft. What is the weir overflow rate (in gallons per day per foot) if the flow over the weir is 2.25 MGD?

Solution:

$$WOR = \frac{2250000\ gpd}{90\ ft}$$

$$= 25000\ gpd/ft$$

Problem 4.20

A sedimentation tank has a total of 210 ft of weir which the water flows over. What is the weir overflow rate (in gallons per day per foot) when the flow is 2.4 MGD?

Solution:

$$WOR = \frac{22400000\ gpd}{210\ ft} = 11,429\ gpd/ft$$

Problem 4.21

The diameter of the weir in a circular clarifier is 120 ft. The flow is 6.12 MGD. What is the weir overflow rate (in gallons per day per foot)?

Solution:

$$WOR = \frac{6120000\ gpd}{(3.14)(120\ ft)}$$

$$= 16,242.04\ gpd/ft$$

Problem 4.22

A tank has a diameter of 48.8 ft. What is the gallons per day per foot of weir overflow when the tank receives 1,955,000 gpd?

Solution:

$$WOR = \frac{1955000 \text{ gpd}}{(3.14)(48.8 \text{ ft})}$$

$$= 12758.6 \text{ gpd/ft}$$

Problem 4.23

The flow rate to a particular clarifier is 530 gpm, and the tank has a length of 32 ft and a width of 18 ft. What is the gallons per day per foot of weir?

Solution:

$$\frac{530 \text{ gpm}}{\text{min}} \frac{1440 \text{ min}}{\text{day}} = 763200 \text{ gpd}$$

$$WOR = \frac{763200 \text{ gpd}}{2(32 \text{ ft}) + 2(18 \text{ ft})} = 7632 \text{ gpd/ft}$$

Problem 4.24

The weir in a basin measures 32 ft by 14 ft. What is the weir overflow rate (in gallons per day per foot) when the flow is 1,096,000 gpd?

Solution:

$$WOR = \frac{1096000 \text{ gpd}}{2(32 \text{ ft}) + 2(14 \text{ ft})} = 11913.04 \text{ gpd/ft}$$

Problem 4.25

What is the weir overflow rate of a clarifier that is 52 ft, 4 in by 46 ft, 3 in and has an influent flow of 1.88 MGD?

Solution:

$$4/12 = 0.3333 \text{ ft}$$
$$3/12 = 0.25 \text{ ft}$$

$$WOR = \frac{1880000 \text{ gpd}}{2(52.3333 \text{ ft}) + 2(46.25 \text{ ft})} = 9534.9 \text{ gpd/ft}$$

Problem 4.26

A tank has a length of 110 ft, a width of 26 ft, and a depth of 16 ft. What is the volume in gallons?

Solution:

$$\text{Vol, gal} = (\text{L, ft}) (\text{W, ft}) (\text{D, ft}) (7.48 \text{ gal/ft}^3)$$
$$= (110 \text{ ft})(26 \text{ ft})(16 \text{ ft})(7.48 \text{ gal/ft}^3)$$
$$= 342,284.8 \text{ gal}$$

Problem 4.27

A clarifier has a diameter of 80 ft and a depth of 10 ft. What is the volume of the clarifier in gallons?

Solution:

$$\text{Vol, gal} = (0.785)(\text{diameter, ft})^2 (\text{depth, ft})(7.48 \text{ gal/ft}^3)$$
$$= (0.785)(80 \text{ ft})(80 \text{ ft})(10 \text{ ft})(7.48 \text{ gal/ft}^3)$$
$$= 375795.2 \text{ gal}$$

Problem 4.28

A circular clarifier handles a flow of 0.8 MGD. The clarifier is 60 ft in diameter and 9 ft deep. What is the detention time in hours?

Solution:

$$DT = \frac{(\text{Vol, gal})((24 \text{ hr/d})}{\text{flow, gpd}}$$
$$= \frac{(0.785)(60 \text{ ft})(60 \text{ ft})(9 \text{ ft})(7.48 \text{ gal/ft}^3)(24 \text{ hr/d})}{800,000 \text{ gpd}} = 5.71 \text{ hrs}$$

Problem 4.29

A clarifier is 72 ft long, 26 ft wide, and 12 ft deep. If the daily flow is 2,760,000 gpd, what is the detention time (in hours) in the basin?

Solution:

$$DT = \frac{(72 \text{ ft})(26 \text{ ft})(12 \text{ ft})(7.48 \text{ gal/ft}^3)(24 \text{ hr/d})}{2,780000 \text{ gpd}} = 1.45 \text{ hrs}$$

Problem 4.30

What is the detention time in hours of a circular clarifier that receives a flow of 3,200 gpm and the clarifier is 60 ft in diameter and 10 ft deep?

Solution:

$$\frac{3200 \text{ gal}}{\text{min}} \frac{1440 \text{ min}}{\text{day}} = 4608000 \text{ gpd}$$

$$DT = \frac{(0.785)(60 \text{ ft})(60 \text{ ft})(10 \text{ ft})(7.48)(24)}{4608000 \text{ gpd}} = 1.10 \text{ hrs}$$

Problem 4.31

A sedimentation tank is 62 ft long, 10 ft wide, and has water to a depth of 10 ft. If the flow to the tank is 21,500 gph, what is the detention time in hours?

Solution:

$$\frac{21{,}500 \text{ gal}}{\text{hr}} \frac{24 \text{ hr}}{\text{day}} = 516000 \text{ gpd}$$

$$DT = \frac{(62 \text{ ft})(10 \text{ ft})(10 \text{ ft})(7.48 \text{ gal/ft}^3)(24 \text{ hr/d})}{516000 \text{ gpd}} = 2.16 \text{ hr}$$

Problem 4.32

A circular clarifier receives a flow of 900 gpm. If it has a diameter of 50 ft and a water depth of 8 ft, what is the detention time in hours?

Solution:

$$\frac{900 \text{ gpm}}{\text{min}} \frac{1440 \text{ min}}{\text{day}} = 1296000 \text{ gpd}$$

$$DT = \frac{(50 \text{ ft})(50 \text{ ft})(0.785)(8 \text{ ft})(7.48 \text{ gal/ft}^3)(24 \text{ hr/d})}{1296000 \text{ gpd}} = 2.17 \text{ hr}$$

Problem 4.33

A clear well is 60 ft long, 15 ft wide, and has water to a depth of 6 ft. If the daily flow is 690 gpm, what is the detention time in minutes?

Solution:

$$\frac{690 \text{ gpm}}{\text{min}} \frac{1440 \text{ min}}{\text{day}} = 993600 \text{ gpd}$$

$$DT, \text{hr} = \frac{(60 \text{ ft})(15 \text{ ft})(6 \text{ ft})(7.85 \text{ gal/ft}^3)(24 \text{ hr/d})}{993600} = 1.02 \text{ hr}$$
$$= (1.02 \text{ hr})(60 \text{ min/hr}) = 61.4 \text{ min}$$

Problem 4.34

The flow to a sedimentation tank is 4.10 MGD. If the tank is 70 ft long and 30 ft wide, what is the surface overflow rate (in gallons per day per square foot)?

Solution:

$$SOR = \frac{4100000 \text{ gpd}}{(70 \text{ ft})(30 \text{ ft})} = 1952.38 \text{ gpd/ft}^2$$

Problem 4.35

The diameter of the weir in a circular clarifier is 120 ft. The flow is 5.50 MGD. What is the weir overflow rate (in gallons per day per foot)?

Solution:

$$\text{WOR} = \frac{5500000 \text{ gpd}}{(3.14)(120 \text{ ft})} = 14596.6 \text{ gpd/ft}$$

Problem 4.36

A rectangular clarifier handles a flow of 3.42 MGD. The clarifier is 60 ft long, 30 ft wide, and 22 ft deep. What is the detention time in minutes?

Solution:

$$\text{DT} = \frac{(60 \text{ ft})(30 \text{ ft})(22 \text{ ft})(7.48 \text{ gal/ft}^3)(24 \text{ hr/day})(60 \text{ min/hr})}{3420000 \text{ gpd}}$$

$$= 124.72 \text{ min}$$

Problem 4.37

A circular clarifier receives a flow of 3,466.4 gpm. What is the detention time in the clarifier (in hours)? The clarifier has a diameter of 63 ft and a depth of 20 ft.

Solution:

$$\text{DT} = \frac{(0.785)(63 \text{ ft})(63 \text{ ft})(20 \text{ ft})(7.48 \text{ gal/ft}^3)(24\text{hr/d})}{(3,466.4 \text{ gal/min})(1440 \text{ min/day})}$$

$$= 2.24 \text{ hours}$$

5 Filtration Calculations

5.1 WATER FILTRATION

Water filtration is a physical process of separating suspended and colloidal particles from waste by passing the water through a granular material. The process of filtration involves straining, settling, and adsorption. As floc passes into the filter, the spaces between the filter grains become clogged, reducing this opening and increasing removal. Some material is removed merely because it settles on a media grain. One of the most important processes is adsorption of the floc onto the surface of individual filter grains. In addition to removing silt and sediment, flock, algae, insect larvae, and any other large elements, filtration also contributes to the removal of bacteria and protozoans, such as *Giardia lamblia* and *Cryptosporidium*. Some filtration processes are also used for iron and manganese removal.

The Surface Water Treatment Rule (SWTR) specifies four filtration technologies, although SWTR also allows the use of alternate filtration technologies, for example, cartridge filters. These include slow sand filtration (see Figure 5.1), rapid sand filtration, pressure filtration, diatomaceous earth filtration, and direct filtration. Of these, all but rapid sand filtration is commonly employed in small water systems that use filtration. Each type of filtration system has advantages and disadvantages. Regardless of the type of filter, however, filtration involves the processes of *straining* (where particles are captured in the small spaces between filter media grains), *sedimentation* (where the particles land on top of the grains and stay there), and *adsorption* (where a chemical attraction occurs between the particles and the surface of the media grains.

5.2 FLOW RATE THROUGH A FILTER (GPM)

Flow rate in gallons per minute through a filter can be determined by simply converting the gallons per day flow rate, as indicated on the flow meter. The flow rate (gallons per minute) can be calculated by taking the meter flow rate (gallons per day) and dividing by 1,440 min/day, as shown in equation 5.1.

$$\text{Flow Rate, gpm} = \frac{\text{Flow Rate, gpd}}{1440 \text{ min/day}} \tag{5.1}$$

Example 5.1

Problem:

The flow rate through a filter is 4.25 MGD. What is this flow rate expressed as gallons per minute?

DOI: 10.1201/9781003354307-5

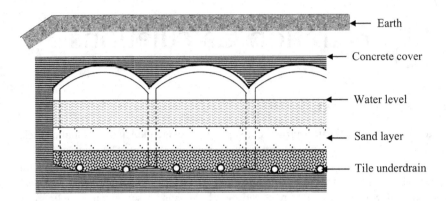

FIGURE 5.1   Slow sand filter.

Solution:

$$\text{Flow Rate, gpm} = \frac{4.25\ \text{MGD}}{1440\ \text{min/day}}$$

$$= \frac{4,250,000\ \text{gpd}}{1440\ \text{min/day}}$$

$$= 2951\ \text{gpm}$$

Example 5.2

Problem:

During a 70 hr filter run, a total of 22.4 MG of water is filtered. What is the average flow rate through the filter in gallons per minute during this filter run?

Solution:

$$\text{Flow Rate, gpm} = \frac{\text{Total Gallons Produced}}{\text{Filter Run, min}}$$

$$= \frac{22,400,000\ \text{gal}}{(70\ \text{hrs})(60\ \text{min/hr})}$$

$$= 5333\ \text{gpm}$$

Example 5.3

Problem:

At an average flow rate of 4,000 gpm, how long a filter run (in hours) would be required to produce 25 MG of filtered water.

Solution:

Write the equation as usual, filling in known data:

$$\text{Flow Rate, gpm} = \frac{\text{Total Gallons Produced}}{\text{Filter Run, min}}$$

$$4000\,\text{gpm} = \frac{25,000,000\,\text{gal}}{(x\,\text{hrs})(60\,\text{min/hr})}$$

Then solve for *x:*

$$= \frac{25,000,000\,\text{gal}}{(4000)(60)}$$

$$= 104\ \text{hrs}$$

Example 5.4

Problem:

A filter box is 20 ft × 30 ft (including the sand area). If the influent valve is shut, the water drops 3 in/min. What is the rate of filtration in million gallons per day?

Solution:

Given:

$$\text{Filter Box} = 20\ \text{ft} \times 30\ \text{ft}$$

$$\text{Water drops} = 3.0\ \text{inches per minute}$$

Find the volume of water passing through the filter:

$$\text{Volume} = \text{Area} \times \text{Height}$$

$$\text{Area} = \text{Width} \times \text{Length}$$

Note: The best way to perform calculations for this type of problem is step-by-step, breaking down the problem into what is given and what is to be found.

Step 1:

$$\text{Area} = 20\ \text{ft} \times 30\ \text{ft} = 600\ \text{ft}^2$$

Convert 3 in into feet. Divide 3 by 12 to find feet.

$$3.0/12 = 0.25\ \text{feet}$$

$$\text{Volume} = 600\ \text{ft}^2 \times 0.25\ \text{ft}$$

$$= 150\ \text{ft}^3\text{of water passing through the filter in one minute}$$

Step 2: Convert cubic feet to gallons.

$$150\ \text{ft}^3 \times 7.48\ \text{gal/ft}^3 = 1,122\ \text{gal/min}$$

Step 3: The problem asks for the rate of filtration in million gallons per day. To find million gallons per day, multiply the number of gallons per minute by the number of minutes per day.

$$1{,}122 \text{ gal/min} \times 1{,}440 \text{ min/day} = 1.62 \text{ MGD}$$

Example 5.5

Problem:

The influent valve to a filter is closed for 5 min. During this time, the water level in the filter drops 0.8 ft (10 in). If the filter is 45 ft long and 15 ft wide, what is the gallons per minute flow rate through the filter? Water drop equals 0.16 ft/min.

Solution:

First, calculate cubic feet per minute flow rate using the Q = AV equation:

$$Q, \text{cfm} = (\text{Length, ft}) \ (\text{Width, ft}) \ (\text{Drop Velocity, ft/min})$$
$$= (45 \text{ ft})(15 \text{ ft})(0.16 \text{ ft/min})$$
$$= 108 \text{ cfm}$$

Then convert cubic feet per minute flow rate to gallons per minute flow rate:

$$(108 \text{ cfm})(7.48 \text{ gal/cu ft}) = 808 \text{ gpm}$$

5.3 FILTRATION RATE

One measure of filter production is filtration rate (generally ranges from 2 to 10 gpm/ sq ft). Along with filter run time, it provides valuable information for operation of filters. It is the gallons per minute of water filtered through each square foot of filter area. Filtration rate is determined using equation 5.2.

$$\text{Filtration Rate, gpm/sq ft} = \frac{\text{Flow Rate, gpm}}{\text{Filter Surface Area, sq ft}} \qquad (5.2)$$

Example 5.6

Problem:

A filter 18 ft by 22 ft receives a flow of 1,750 gpm. What is the filtration rate in gallons per minute per square foot?

Solution:

$$\text{Filtration Rate} = \frac{\text{Flow Rate, gpm}}{\text{Filter Surface Area, sq ft}}$$
$$= \frac{1750 \text{ gpm}}{(18 \text{ ft})(22 \text{ ft})}$$
$$= 4.4 \text{ gpm/sq ft}$$

Example 5.7

Problem:

A filter 28 ft long and 18 ft wide treats a flow of 3.5 MGD. What is the filtration rate in gallons per minute per square foot?

Solution:

$$\text{Flow Rate} = \frac{3,500,000 \text{ gpd}}{1440 \text{ min/day}} = 2431 \text{ gpm}$$

$$\text{Filtration Rate, gpm/sq ft} = \frac{\text{Flow Rate, gpm}}{\text{Filter Surface Area, sq ft}}$$

$$= \frac{2431 \text{ gpm}}{(28 \text{ ft})\,(18 \text{ ft})}$$

$$= 4.8 \text{ gpm/sq ft}$$

Example 5.8

Problem:

A filter 45 ft long and 20 ft wide produces a total of 18 MG during a 76 hr filter run. What is the average filtration rate in gallons per minute per square foot for this filter run?

Solution:

First, calculate the gallons per minute flow rate through the filter:

$$\text{Flow Rate, gpm} = \frac{\text{Total Gallons Produced}}{\text{Filter Run, min}}$$

$$= \frac{18,000,000 \text{ gal}}{(76 \text{ hrs})\,(60 \text{ min/hr})}$$

$$= 3947 \text{ gpm}$$

Then calculate filtration rate:

$$\text{Filtration Rate} = \frac{\text{Flow Rate, gpm}}{\text{Filter Area, sq ft}}$$

$$= \frac{3947 \text{ gpm}}{(45 \text{ ft})\,(20 \text{ ft})}$$

$$= 4.4 \text{ gpm/sq ft}$$

Example 5.9

Problem:

A filter is 40 ft long and 20 ft wide. During a test of flow rate, the influent valve to the filter is closed for 6 min. The water level drop during this period is 16 in. What is the filtration rate for the filter in gallons per minute per square foot?

Solution:

First, calculate gallons per minute flow rate using the Q = AV equation:

$$Q, gpm = (Length, ft) \ (Width, ft) \ (Drop \ Velocity, ft/min) \ (7.48 \ gal/cu \ ft)$$

$$= \frac{(40 \ ft) \ (20 \ ft) \ (1.33 \ ft) \ (7.48 \ gal/cu \ ft)}{6 \ min}$$

$$= 1316 \ gpm$$

Then calculate filtration rate:

$$Filtration \ Rate = \frac{Flow \ Rate, \ gpm}{Filter \ Area, \ sq \ ft}$$

$$= \frac{1316 \ gpm}{(40 \ ft) \ (20 \ ft)}$$

$$= 1.657 \ or \ 1.7 \ gpm/sq \ ft$$

5.4   UNIT FILTER RUN VOLUME (UFRV)

The unit filter run volume (UFRV) calculation indicates the total gallons passing through each square foot of filter surface area during an entire filter run. This calculation is used to compare and evaluate filter runs. UFRVs are usually at least 5,000 gal/sq ft and generally in the range of 10,000 gpd/sq ft. The UFRV value will begin to decline as the performance of the filter begins to deteriorate. The equation to be used in these calculations is shown here:

$$UFRV = \frac{Total \ Gallons \ Filtered}{Filter \ Surface \ Area, \ sq \ ft} \qquad (5.3)$$

Example 5.10

Problem:

The total water filtered during a filter run (between backwashes) is 2,220,000 gal. If the filter is 18 ft by 18 ft, what is the unit filter run volume (UFRV) in gallons per square foot?

Solution:

$$UFRV = \frac{Total \ Gallons \ Filtered}{Filter \ Surface \ Area, \ sq \ ft}$$

$$= \frac{2,220,000 \ gal}{(18 \ ft) \ (18 \ ft)}$$

$$= 6852 \ gal/sq \ ft$$

Example 5.11

Problem:

The total water filtered during a filter run is 4,850,000 gal. If the filter is 28 ft by 18 ft, what is the unit filter run volume in gallons per square foot?

Solution:

$$UFRV = \frac{\text{Total Gallons Filtered}}{\text{Filter Surface Area, sq ft}}$$

$$= \frac{4,850,000 \text{ gal}}{(28 \text{ ft})(18 \text{ ft})}$$

$$= 9623 \text{ gal/sq ft}$$

Equation 5.3 can be modified, as shown in equation 5.4, to calculate the unit filter run volume given filtration rate and filter run data.

$$UFRV = (\text{Filtration Rate, gpm, sq ft})(\text{Filter Run Time, min}) \qquad (5.4)$$

Example 5.12

Problem:

The average filtration rate for a filter was determined to be 2.0 gpm/sq ft. If the filter run time was 4,250 min, what was the unit filter run volume in gallons per square foot?

Solution:

$$UFRV = (\text{Filtration Rate, gpm/sq ft})(\text{Filter Run Time, min})$$

$$= 8500 \text{ gal/sq ft}$$

The problem indicates that at an average filtration rate of 2.0 gal entering each square foot of filter each minute, the total gallons entering during the total filter run is 4,250 times that amount.

Example 5.13

Problem:

The average filtration rate during a particular filter run was determined to be 3.2 gpm/sq ft. If the filter run time was 61 hr, what was the UFRV in gallons per square foot for the filter run?

Solution:

$$UFRV = (\text{Filtration Rate, gpm/sq ft})(\text{Filter Run, hrs})(60 \text{ min/hr})$$

$$= (3.2 \text{ gpm/sq ft})(61.0 \text{ hrs})(60 \text{ min/hr})$$

$$= 11,712 \text{ gal/sq ft}$$

5.5   BACKWASH RATE

In filter backwashing, one of the most important operational parameters to be determined is the amount of water in gallons required for each backwash. This amount depends on the design of the filter and the quality of the water being filtered. The actual washing typically lasts 5 to 10 min and uses amounts to 1–5% of the flow produced.

Example 5.14

Problem:

A filter has the following dimensions:

$$\text{Length} = 30 \text{ ft}$$
$$\text{Width} = 20 \text{ ft}$$
$$\text{Depth of filter media} = 24 \text{ inches}$$

Assuming a backwash rate of 15 gal/sq ft/min is recommended and 10 min of backwash is required, calculate the amount of water in gallons required for each backwash.

Solution:

Given:

$$\text{Length} = 30 \text{ ft}$$
$$\text{Width} = 20 \text{ ft}$$
$$\text{Depth of filter media} = 24 \text{ inches}$$
$$\text{Rate} = 15 \text{ gal/ft}^2\text{/min}$$

Find the amount of water in gallons required:

Step 1: Area of filter = 30 ft × 20 ft = 600 ft².

Step 2: Gallons of water used per square foot of filter = 15 gal/ft²/min × 10 min = 150 gal/ft².

Step 3: Gallons required = 150 gal/ft² × 600 ft² = 90,000 gals required for backwash. Typically, backwash rates will range from 10 to 25 gpm/sq ft. The backwash rate is determined by using equation 5.5.

$$\text{Backwash} = \frac{\text{Flow Rate, gpm}}{\text{Filter Area, sq ft}} \tag{5.5}$$

Example 5.15

Problem:

A filter 30 ft by 10 ft has a backwash rate of 3,120 gpm. What is the backwash rate in gallons per minute per square foot?

Solution:

$$\text{Backwash Rate, gpm/sq ft} = \frac{\text{Flow Rate, gpm}}{\text{Filter Area, sq ft}}$$
$$= \frac{3120 \text{ gpm}}{(30 \text{ ft})\,(10 \text{ ft})}$$
$$= 10.4 \text{ gpm/sq ft}$$

Example 5.16

Problem:

A filter 20 ft long and 20 ft wide has a backwash flow rate of 4.85 MGD. What is the filter backwash rate in gallons per minute per square foot?

Solution:

$$\text{Backwash Rate} = \frac{\text{Flow Rate, gpm}}{\text{Filter Area, sq ft}}$$

$$= \frac{4,850,000 \text{ gpd}}{1440 \text{ min/day}}$$

$$= 3368 \text{ gpm}$$

$$= \frac{3368 \text{ gpm}}{(20 \text{ ft})(20 \text{ ft})}$$

$$= 8.42 \text{ gpm/sq ft}$$

5.5.1 BACKWASH RISE RATE

Backwash rate is occasionally measured as the upward velocity of the water during backwashing—expressed as inches per minute rise. To convert from gallons per minute per square foot backwash rate to inches per minute rise rate, use either equation 5.6 or equation 5.7:

$$\text{Backwash Rate, in./min} = \frac{(\text{Backwash Rate, gpm/sq ft})(12 \text{ in./ft})}{7.48 \text{ gal/cu ft}} \qquad (5.6)$$

$$\text{Backwash Rate, in./min} = (\text{Backwash Rate, gpm/sq ft})(1.6) \qquad (5.7)$$

Example 5.17

Problem:

A filter has a backwash rate of 16 gpm/sq ft. What is this backwash rate expressed as inches per minute rise rate?

$$\text{Backwash Rate, in./min} = \frac{(\text{Backwash Rate, gpm/sq ft})(12 \text{ in./ft})}{7.48 \text{ gal/cu ft}}$$

$$= \frac{(16 \text{ gpm/sq ft})(12 \text{ in./ft})}{7.48 \text{ gal/cu ft}}$$

$$= 25.7 \text{ in./min}$$

Example 5.18

Problem:

A filter 22 ft long and 12 ft wide has a backwash rate of 3,260 gpm. What is this backwash rate expressed as inches per minute rise?

Solution:

First, calculate the backwash rate as gallons per minute per square foot:

$$\text{Backwash Rate} = \frac{\text{Flow Rate, gpm}}{\text{Filter Area, sq ft}}$$

$$= \frac{3260 \text{ gpm}}{(22 \text{ ft}) \ (12 \text{ ft})}$$

$$= 12.3 \text{ gpm/sq ft}$$

Then convert gallons per minute per square foot to inches per minute rise rate:

$$= \frac{(12.3 \text{ gpm/sq ft}) \ (12 \text{ in./ft})}{7.48 \text{ gal/cu ft}}$$

$$= 19.7 \text{ in./min}$$

5.6   VOLUME OF BACKWASH WATER REQUIRED (GAL)

To determine the volume of water required for backwashing, we must know both the desired backwash flow rate (in gallons per minute) and the duration of backwash (in minutes):

$$\text{Backwash Water Vol., gal} = (\text{Backwash, gpm}) \ (\text{Duration of Backwash, min}) \quad (5.8)$$

Example 5.19

Problem:

For a backwash flow rate of 9,000 gpm and a total backwash time of 8 min, how many gallons of water will be required for backwashing?

Solution:

$$\text{Backwash Water Vol., gal} = (\text{Backwash, gpm}) \ (\text{Duration of Backwash, min})$$

$$= (9,000 \text{ gpm}) \ (8 \text{ min})$$

$$= 72,000 \text{ gal}$$

Example 5.20

Problem:

How many gallons of water would be required to provide a backwash flow rate of 4,850 gpm for a total of 5 min?

$$\text{Backwash Water Vol., gal} = (\text{Backwash, gpm}) \ (\text{Duration of Backwash, min})$$

$$= (4,850 \text{ gpm}) \ (7 \text{ min})$$

$$= 33,950 \text{ gal}$$

5.7   REQUIRED DEPTH OF BACKWASH WATER TANK (FT)

The required depth of water in the backwash water tank is determined from the volume of water required for backwashing. To make this calculation, simply use equation 5.9.

$$\text{Volume, gal} = (0.785)\ (D^2)\ (\text{Depth, ft})\ (7.48\ \text{gal/cu ft}) \qquad (5.9)$$

Example 5.21

Problem:

The volume of water required for backwashing has been calculated to be 85,000 gal. What is the required depth of water in the backwash water tank to provide this amount of water if the diameter of the tank is 60 ft?

Solution:

Use the volume equation for a cylindrical tank, fill in known data, then solve for x:

$$\text{Volume, gal} = (0.785)\ (D^2)\ (\text{Depth, ft})\ (7.48\ \text{gal/cu ft})$$

$$85,000\ \text{gal} = (0.785)\ (60\ \text{ft})\ (60\ \text{ft})\ (x\ \text{ft})\ (7.48\ \text{gal/cu ft})$$

$$= \frac{85,000}{(0.785)\ (60)\ (60)\ (7.48)}$$

$$x = 4\ \text{ft}$$

Example 5.22

Problem:

A total of 66,000 gal of water will be required for backwashing a filter at a rate of 8,000 gpm for a 9 min period. What depth of water is required if the backwash tank has a diameter of 50 ft?

Solution:

Use the volume equation for cylindrical tanks:

$$\text{Volume, gal} = (0.785)\ (D^2)\ (\text{Depth, ft})\ (7.48\ \text{gal/cu ft})$$

$$66,000\ \text{gal} = (0.785)\ (50\ \text{ft})\ (50\ \text{ft})\ (x\,\text{ft})\ (7.48\ \text{gal/cu ft})$$

$$x = \frac{66,000}{(0.785)\ (50)\ (50)\ (7.48)}$$

$$x = 4.5\ \text{ft}$$

5.8   BACKWASH PUMPING RATE (GPM)

The desired backwash pumping rate (in gallons per minute) for a filter depends on the desired backwash rate in gallons per minute per square foot and the square foot area of the filter. The backwash pumping rate, in gallons per minute, can be determined by using equation 5.20.

$$\text{Backwash Pumping Rate, gpm} = (\text{Desired Backwash Rate, gpm/sq ft})$$
$$(\text{Filter Area, sq ft}) \tag{5.10}$$

Example 5.23

Problem:

A filter is 25 ft long and 20 ft wide. If the desired backwash rate is 22 gpm/sq ft, what backwash pumping rate (in gallons per minute) will be required?

Solution:

The desired backwash flow through each square foot of filter area is 20 gpm. The total gallons per minute flow through the filter is therefore 20 gpm times the entire square foot area of the filter:

$$\text{Backwash Pump. Rate, gpm} = (\text{Desired Backwash Rate, gpm/sq ft})$$
$$(\text{Filter Area, sq ft})$$
$$= 20 \text{ gpm/sq ft}) \ (25 \text{ ft}) \ (20 \text{ ft})$$
$$= 10{,}000 \text{ gpm}$$

Example 5.24

Problem:

The desired backwash pumping rate for a filter is 12 gpm/sq ft. If the filter is 20 ft long and 20 ft wide, what backwash pumping rate (in gallons per minute) will be required?

Solution:

$$\text{Backwash Pumping Rate, gpm} = (\text{Desired Backwash Rate, sq ft})$$
$$(\text{Filter Area, sq ft})$$
$$= (12 \text{ gpm/sq ft}) \ (20 \text{ ft}) \ (20 \text{ ft})$$
$$= 4{,}800 \text{ gpm}$$

5.9   PERCENT PRODUCT WATER USED FOR BACKWATERING

Along with measuring filtration rate and filter run time, another aspect of filter operation that is monitored for filter performance is the percent of product water used for backwashing. The equation for percent of product water used for backwashing calculations used is shown next.

$$Backwash \ Water, \ \% = \frac{Backwash \ Water, \ gal}{Water \ Filtered, \ gal} \times 100 \tag{5.11}$$

Example 5.25

Problem:

A total of 18,100,000 gal of water was filtered during a filter run. If 74,000 gal of this product water were used for backwashing, what percent of the product water was used for backwashing?

Solution:

$$\text{Backwash Water, } \% = \frac{\text{Backwash Water, gal}}{\text{Water Filtered, gal}} \times 100$$

$$= \frac{74,000 \text{ gal}}{18,100,000 \text{ gal}} \times 100$$

$$= 0.4 \%$$

Example 5.26

Problem:

A total of 11,400,000 gal of water is filtered during a filter run. If 48,500 gal of product water are used for backwashing, what percent of the product water is used for backwashing?

Solution:

$$\text{Backwash Water, } \% = \frac{\text{Backwash Water, gal}}{\text{Water Filtered, gal}} \times 100$$

$$= \frac{48,500 \text{ gal}}{11,400,000 \text{ gal}} \times 100$$

$$= 0.43 \% \text{ Backwash Water}$$

5.10 PERCENT MUD BALL VOLUME

Mud balls are heavier deposits of solids near the top surface of the medium that break into pieces during backwash, resulting in spherical accretions (usually less than 12 in in diameter) of floc and sand. The presence of mud balls in the filter media is checked periodically. The principal objection to mud balls is that they diminish the effective filter area. To calculate the percent mud ball volume, we use equation 5.12.

$$\% \text{ Mud Ball Volume} = \frac{\text{Mud Ball Volume, mL}}{\text{Total Sample, Volume, mL}} \times 100 \qquad (5.12)$$

Example 5.27

Problem:

A 3,350 mL sample of filter media was taken for mud ball evaluation. The volume of water in the graduated cylinder rose from 500 mL to 525 mL when mud balls were placed in the cylinder. What is the percent mud ball volume of the sample?

Solution:

First, determine the volume of mud balls in the sample:

$$525 \text{ mL} - 500 \text{ mL} = 25 \text{ mL}$$

Then calculate the percent mud ball volume:

$$= \frac{25 \text{ mL}}{3350 \text{ mL}} \times 100$$

$$= 0.70\%$$

Example 5.28

Problem:

A filter is tested for the presence of mud balls. The mud ball sample has a total sample volume of 680 mL. Five samples were taken from the filter. When the mud balls were placed in 500 mL of water, the water level rose to 565 mL. What is the percent mud ball volume of the sample?

Solution:

The mud ball volume is the volume the water rose:

$$565 \text{ mL} - 500 \text{ mL} = 65 \text{ mL}$$

Because five samples of media were taken, the total sample volume is 5 times the sample volume:

$$(5)(680 \text{ mL}) = 3400 \text{ mL}$$

$$\% \text{ Mud Ball Volume} = \frac{65 \text{ mL}}{3400 \text{ mL}} \times 100$$

$$= 1.9\%$$

5.11 FILTER BED EXPANSION

In addition to backwash rate, it is also important to expand the filter media during the wash to maximize the removal of particles held in the filter or by the media, that is, the efficiency of the filter wash operation depends on the expansion of the sand bed. Bed expansion is determined by measuring the distance from the top of the unexpanded media to a reference point (e.g., top of the filter wall) and from the top of the expanded media to the same reference. A proper back wash rate should expand the filter 20–25%. Percent bed expansion is given by dividing the bed expansion by the total depth of expandable media (i.e., media depth less support gravels) and multiplied by 100, as follows:

expanded measurement = depth to top of media during backwash (inches)
unexpanded measurement = depth to top of media before backwash (inches)
bed expansion = unexpanded measurement (inches) – expanded measurement (inches)

$$\text{Bed Expansion , \%} = \frac{\text{Bed expansion measurement (inches)}}{\text{Total depth of expandable media (inches)}} \times 100 \qquad (5.13)$$

Example 5.29[1]

Problem:

The backwashing practices for a filter with 30 in of anthracite and sand are being evaluated. While at rest, the distance from the top of the media to the concrete floor surrounding the top of the filter is measured to be 41 in. After the backwash has been started and the maximum backwash rate is achieved, a probe containing a white disk is slowly lowered into the filter bed until anthracite is observed on the disk. The distance from the expanded media to the concrete floor is measured to be 34 in. What is the percent bed expansion?

Solution:

Given:

> Unexpanded measurement: 41 in
> Expanded measurement: 34.5 in
> Bed expansion: 6.5
> Bed expansion (percent): (6.5 in/30 in) × 100 = 22%

5.12   FILTER LOADING RATE

Filter loading rate is the flow rate of water applied to the unit area of the filter. It is the same value as the flow velocity approaching the filter surface and can be determined by using equation 5.14.

$$u = Q/A \qquad (5.14)$$

Where:
u = loading rate, $m^3/(m^2\ d)$ or gpm/ft^2
Q = flow rate, m^3/d or ft^3/d of gpm
A = surface area of filter, in square meter or square foot

Filters are classified as slow sand filters, rapid sand filters, and high-rate sand filters on basis of loading rate. Typically, the loading rate for rapid sand filters is 120 m3/$(m^2\ d)(83\ L/(m^2\ min)$ or 2 gal/min/ft². The loading rate may be up to five times this rate for high-rate filters.

Example 5.30

Problem:

A sanitation district is to install rapid sand filters downstream of the clarifiers. The design loading rate is selected to be 150 m³/m². The design capacity of the

---

[1] USEPA, 1999. *Individual Filter Self Assessment: EPA Guidance Manual, Turbidity Provisions.* Washington, DC: Environment Protection Agency, pp. 5–12.

waterworks is 0.30 m³/s (6.8 MGD). The maximum surface per filter is limited to 45 m². Design the number and size of filters and calculate the normal filtration rate.

Solution:

Step 1: Determine the total surface area required.

$$A = \frac{Q}{u} = \frac{0.30 \text{ m}^3/\text{sec } (85{,}400 \text{ sec/day})}{150 \text{ m}^3/\text{m}^2\text{day}}$$

$$= \frac{25920}{150}$$

$$= 173 \text{ m}^2$$

Step 2: Determine the number of filters.

$$\frac{173 \text{ m}^2}{45 \text{ m}} \, 3.8$$

Select four meters.

The surface area (*a*) for each filter is:

$$a = 173 \text{ m}^2/4 = 43.25 \text{ m}^2$$

We can use 6 m × 7 m or 6.4 m × 7 m or 6.42 m × 7 m.

Step 3: If a 6 m × 7 m filter is installed, the normal filtration rate is:

$$u = \frac{Q}{A} = \frac{0.30 \text{ m}^3/\text{s} \times 86{,}400 \text{ s/d}}{4 \times 6 \text{ m} \times 7 \text{ m}}$$

$$= 154.3 \text{ m}^3/\left(\text{m}^2\text{d}\right)$$

5.13   FILTER MEDIUM SIZE

Filter medium grain size has an important effect on the filtration efficiency and on backwashing requirements for the medium. The actual medium selected is typically determined by performing a grain size distribution analysis—sieve size and percentage passing by weight relationships are plotted on logarithmic probability paper. The most common parameters used in the United States to characterize the filter medium are effective size (ES) and uniformity coefficient (UC) of medium size distribution. The ES is that grain size for which 10% of the grains are smaller by weight; it is often abbreviated by d^{10}. The UC is the ratio of the 60 percentile (d^{60}) to the 10 percentile. The 90 percentile, d_{90}, is the size for which 90% of the grains are smaller by weight. The d_{90} size is used for computing the required filter backwash rate for a filter medium.

Values of d_{10}, d_{60}, and d_{90} can be read from an actual sieve analysis curve. If such a curve is not available, and if a linear log probability plot is assumed, the values can be interrelated by equation 5.15 (Cleasby, 1990).

$$d_{90} = d_{10}\left(10^{1.67 \log \text{UC}}\right) \tag{5.15}$$

Example 5.31

Problem:

A sieve analysis curve of a typical filter sand gives $d_{10} = 0.52$ mm and $d_{60} = 0.70$ mm. What are its uniformity coefficient and d_{90}?

Solution:

Step 1: $UC = d_{60} / d_{10} = 0.70$ mm $/ 0.52$ mm .

$\qquad\qquad = 1.35$

Step 2:   Find d_{90} using equation 5.15.

$$D_{90} = d_{10} \left(10^{1.67 \ \log \ UC} \right)$$

$$= 0.52 \ \text{mm} \left(10^{1.67 \ \log \ 1.35} \right)$$

$$= 0.52 \ \text{mm} \left(10^{0.218} \right)$$

$$= 0.86 \ \text{mm}$$

5.14 MIXED MEDIA

Recently, an innovation in filtering systems has offered a significant improvement and economic advantage to rapid rate filtration: this is the mixed media filter bed. Mixed media filter beds offer specific advantages in specific circumstances but will give excellent operating results at a filtering rate of 5 gal/sq ft/min. Moreover, the mixed media filtering unit is more tolerant to handling higher turbidities in the settled water. For improved process performance, activated carbon or anthracite is added on the top of the sand bed. The approximate specific gravity (*s*) of ilmenite (Chavara, <60% TiO_2), silica sand, anthracite, and water is 4.2, 2.6, 1.5, and 1.0, respectively. The economic advantage of the mixed bed media filter is based upon filter area; it will safely produce 2 1/2 times as much filtered water as a rapid sand filter.

When settling velocities are equal, the particle sizes for media of different specific gravity can be computed by using equation 5.16.

$$\frac{d_1}{d_2} = \left(\frac{s_2 - s}{s_1 - s} \right)^{2/3} \qquad\qquad (5.16)$$

Where:

d_1, d_2 = diameter of particles 1 and 2 and water, respectively
s_1, s_2 = specific gravity of particles 1, 2, and water, respectively

Example 5.32

Problem:

Estimate the particle size of ilmenite sand (specific gravity = 4.2) that has the same settling velocity of silica sand 0.60 mm in known diameter (specific gravity = 2.6).

Solution:

Find the diameter on ilmenite sand by using equation 5.16.

$$d = (0.6 \text{ mm}) \left(\frac{2.6 - 1}{4.2 - 1} \right)^{2/3}$$

$$= 0.38 \text{mm settling size}$$

5.15 HEAD LOSS FOR FIXED-BED FLOW

When water is pumped upward through a bed of fine particles at a very low flow rate, the water percolates through the pores (void spaces) without disturbing the bed. This is a fixed-bed process. The head loss (pressure drop) through a clean granular media filter is generally less than 0.9 m (3 ft). With the accumulation of impurities, head loss gradually increases until the filter is backwashed. The Kozeny equation, shown next, is typically used for calculating head loss through a clean fixed-bed flow filter.

$$\frac{h}{L} = \frac{k\mu(1-\varepsilon)^2}{gp\varepsilon^3} \left(\frac{A}{V} \right)^2 u \tag{5.17}$$

Where:

h = head loss in filter depth in liter, meter, or feet
k = dimensionless Kozeny constant; 5 for sieve openings, 6 for size of separation
g = acceleration of gravity, 9.81 m/s or 32.2 ft/s
μ = absolute viscosity of water, N s/m^2 or lb s/ft^2
p = density of water, kg/m^3 or lb/ft^3
ε = porosity, dimensionless
A/V = grain surface area per unit volume of grain
= *specific surface S (or shape factor = 6.0–7.7)*
= 6/d for spheres
= $6/\psi d_{eq}$ for irregular grains
ψ = grain sphericity or shape factor
d_{eq} = grain diameter of spheres of equal volume
u = filtration (superficial) velocity, m/s or fps

Example 5.33

Problem:

A dual medium filter is composed of 0.3 m anthracite (mean size of 2.0 mm) that is placed over a 0.6 m layer of sand (mean size 0.7 mm) with a filtration rate of 9.78 m/h. Assume the grain sphericity is ψ = 0.75 and a porosity for both is 0.42. Although normally taken from the appropriate table at 15°C, we provide the head loss data of the filter at 1.131 × 10^{-6} m^2 sec.

Solution:

Step 1: Determine head loss through anthracite layer using the Kozeny equation (equation (5.17).

$$\frac{h}{L} = \frac{k\mu(1-\varepsilon)^2}{gp\varepsilon^3}\left(\frac{A}{V}\right)^2 u$$

Where:

$k = 6$

$g = 9.81$ m/s^2

$\mu p = v = 1.131 \times 10^{-6}$ m^2 s (from the appropriate table)

$\varepsilon = 0.40$

$A/V = 6/0.75d = 8/d = 8/0.002$

$u = 9.78$ m/h $= 0.00272$ m/s

$L = 0.3$ m

Then:

$$h = 6\times\frac{1.131\times10^{-6}}{9.81}\times\frac{(1-0.42)^2}{0.42^3}\times\left(\frac{8}{0.002}\right)^2(0.00272)\,(0.2)$$

$$= 0.0410 \text{ m}$$

Step 2: Compute the head loss passing through the sand. Use data in step 1, except insert:

$k = 5$

$d = 0.0007$ m

$L = 0.6$ m

$$h = 5\times\frac{1.131\times10^{-6}}{9.81}\times\frac{0.58^2}{0.42^3}\times\left(\frac{8}{d}\right)^2(0.00272)\,(0.4)$$

$$= 0.5579 \text{ m}$$

Step 3: Compute total head loss.

$$h = 0.0410 \text{ m} + 0.5579 \text{ m}$$

$$= 0.599 \text{ m}$$

5.16 HEAD LOSS THROUGH A FLUIDIZED BED

If the upward water flow rate through a filter bed is very large, the bed mobilizes pneumatically and may be swept out of the process vessel. At an intermediate flow rate, the bed expands and is in what we call an *expanded* state. In the fixed bed, the particles are in direct contact with each other, supporting each other's weight. In the expanded bed, the particles have a mean free distance between particles and the drag force of the water supports the particles. The expanded bed has some of the properties of the water (i.e., of a fluid) and is called a fluidized bed (Chase, 2002). Simply, *fluidization* is defined as upward flow through a granular filter bed at sufficient velocity to suspend the grains in the water. Minimum fluidizing velocity (U_{mf}) is the superficial fluid velocity needed to start fluidization; it is important in determining the required minimum backwashing flow rate. Wen and Yu proposed the U_{mf} equation, including the near constants (over a wide range of particles) 33.7 and 0.0408, but excluding porosity of fluidization and shape factor (Wen & Yu, 1966):

$$U_{mf} = \frac{\mu}{pd_{eq}}\left(1135.69 + 0.0408G_n\right)^{0.5} - \frac{33.7\mu}{pd_{eq}} \qquad (5.18)$$

Where:
   μ = absolute viscosity of water, N s/m² or lb s/ft²
   p = density of water, kg/m³ or lb/ft³
   $d_{eq} = d_{90}$ sieve size is used instead of d_{eq}
   G_n = Galileo number

$$= d^3{}_{eq}P\left(p_s - p\right)g / \mu^2 \qquad (5.19)$$

Note: Based on the studies of Cleasby and Fan (1981), we use a safety factor of 1.3 to ensure adequate movement of the grains.

Example 5.34

Problem:

Estimate the minimum fluidized velocity and backwash rate for the sand filter. The d_{90} size of sand is 0.90 mm. The density of sand is 2.68 g/cm³.

Solution:

Step 1: Compute the Galileo number.
   From given data and the applicable table, at 15°C.
   $p = 0.999$ g/cm³
   $\mu = 0.0113$ N s/m² = 0.00113 kg/ms = 0.0113 g/cm s
   $\mu p = 0.0113$ cm²/s
   $g = 981$ cm/s²
   $d = 0.90$ cm
   $p_s = 2.68$ g/cm³
   Using equation 5.19:

$$G_n = d^3{}_{eq}P\left(p_s - p\right)g / m^2$$
$$= (0.090)^3 (0.999)\ (2.68 - 0.999)\ (981)/(0.0113)^2$$
$$= 9405$$

Step 2: Compute U_{mf} using equation 5.19.

$$U_{mf} = \frac{0.0113}{0.999 \times 0.090}\left(1135.69 + 0.0408 \times 9405\right)^{0.5} - \frac{33.7 \times 0.0113}{0.999 \times 0.090}$$
$$= 0.660 \text{ cm/s}$$

Step 3: Compute backwash rate. Apply a safety factor of 1.3 to U_{mf} as backwash rate:

$$\text{Backwash rate} = 1.3 \times 0.660 \text{ cm/s} = 0.858 \text{ cm/s}$$

$$0.858 \frac{\text{cm}^3}{\text{cm}^2\text{s}} \times \frac{\text{L}}{1000 \text{ cm}^3} \times \frac{1}{3.785} \times \frac{\text{gal}}{\text{L}} \times 929 \times \frac{\text{cm}^2}{\text{ft}^2} \times \frac{60\text{s}}{\text{min}}$$

$$= 12.6 \text{ pgm/ft}^2$$

5.17 HORIZONTAL WASHWATER TROUGHS

Wastewater troughs are used to collect backwash water as well as to distribute influent water during the initial stages of filtration. Washwater troughs are normally placed above the filter media, specifically in the United States. Proper placement of these troughs is very important to ensure that the filter media is not carried into the troughs during the backwash and removed from the filter. These backwash troughs are constructed from concrete, plastic, fiberglass, or other corrosion-resistant materials. The total rate of discharge in a rectangular trough with free flow can be calculated by using equation 5.20.

$$Q = Cwh^{1.5} \qquad\qquad (5.20)$$

Where:
  Q = flow rate, in cubis feet per second
  C = constant (2.49)
  w = trough width, in feet
  h = maximum water depth in trough, in feet

Example 5.35

Problem:
Troughs are 18 ft long, 18 in wide, and 8 ft to the center, with a horizontal flat bottom. The backwash rate is 24 in/min. Estimate (a) the water depth of the troughs with free flow into the gullet and (b) the distance between the top of the troughs and the 30 in sand bed. Assume 40% expansion and 6 in of freeboard in the troughs and 6 in of thickness.

Solution:
Step 1: Estimate the maximum water depth (h) in trough:

$$v = 24 \text{ in/min} = 2 \text{ ft/60s} = 1/30 \text{ fps}$$

$$A = 18 \text{ ft} \times 8 \text{ ft} = 144 \text{ ft}^2$$

$$Q = VA = 144/30 \text{ cfs}$$

$$= 4.8 \text{ cfs}$$

Using equation 5.20:

$$Q = 2.49 \, wh^{1.5}, \; w = 1.5 \text{ ft}$$
$$h = (Q / 2.49 \, w)^{2/3}$$
$$= \left[4.8 / (2.49 \times 1.5) \right]^{2/3}$$
$$= 1.18 \text{ ft (or approximately 14 in.} = 1.17 \text{ ft)}$$

Step 2: Determine the distance (y) between the sand bed surface and the top troughs.

$$\text{Freeboard} = 6 \text{ in.} = 0.5 \text{ ft}$$
$$\text{Thickness} = 8 \text{ in} = 0.67 \text{ ft (the bottom of trough)}$$
$$y = 2.5 \text{ ft} \times 0.4 + 1.17 \text{ ft} + 0.5 \text{ ft} + 0.5 \text{ ft}$$
$$= 3.2 \text{ ft}$$

5.18 FILTER EFFICIENCY

Water treatment *filter efficiency* is defined as the effective filter rate divided by the operation filtration rate, as shown in equation 5.21 (AWWA & ASCE, 1998).

$$E = \frac{R_e}{R_o} = \frac{UFRV - UBWU}{UFRV} \tag{5.21}$$

Where:
E = filter efficiency, %
R_e = effective filtration rate, gpm/ft^2
R_o = operating filtration rate, gpm/ft^2
$UFRV$ = unit filter run volume, gal/ft^2
$UBWV$ = unit backwash volume, gal/ft^2

Example 5.36

Problem:
A rapid sand filter operates at 3.9 gpm/ft^2 for 48 hr. Upon completion of the filter run, 300 gal/ft^2 of backwash water are used. Find the filter efficiency.

Solution:
Step 1: Calculate operating filtration rate, R_o.

$$R_o = 3.9 \text{ gpm/ft}^2 \times 60 \text{ min/h} \times 48 \text{ h}$$
$$= 11,232 \text{ gal/ft}^2$$

Step 2: Calculate effective filtration rate, R_e.

$$R_e = (11,232 - 300) \text{ gal/ft}^2$$
$$= 10,932 \text{ gal/ft}^2$$

Step 3: Calculate filter efficiency, E, using equation 5.21.

$$E = 10,932/11,232$$
$$= 97.3\%$$

5.19   WATER FILTRATION PRACTICE PROBLEMS

Problem 5.1

A waterworks treats an average of 2.85 MGD. The water is split equally to each of six filters. Each filter basin measures 12 ft wide by 22 ft long and 20 ft deep. Each filter bed measures 8 ft by 16 ft by 14 ft deep.

  a. Determine the daily flow to each of the filters in gallons per minute.

Solution:

$$\frac{2,850,000 \text{ gpd}}{6 \text{ filters}} = 475,000 \text{ gpd}$$

  b. The influent line to filter 5 is closed, while the effluent remains open. Using a hook gauge and a stopwatch, it is noted that the water level in the filter drops 6 in in 60 sec. What is the filtration rate in gallons per minute?

Solution:

$$\text{Vol, gal} = (12 \text{ ft})(22 \text{ ft})(0.5\text{ft})(7.48) = 987.4 \text{ gal}$$
$$\text{Time} = 60 \text{ s}/60 \text{ s/min} = 1 \text{ min}$$
$$\text{Gpm} = \frac{987.4 \text{ gal}}{1 \text{ min}} = 987.4 \text{ gal/min}$$

  c. What is the filtration rate in gallon per minute per filter per square foot of surface area?

Solution:

$$\text{Basin area} = (12 \text{ ft})(22 \text{ ft}) = 264 \text{ ft}^2$$
$$\text{Filter rate} = \frac{987.4 \text{ gpm}}{264 \text{ ft}^2} = 3.74 \text{ gpm/ft}^2$$

  d. A hook gauge was used to determine the rate of rise in the filter bed during the backwash cycle. The water rose 6 in in 30 sec. What is the backwash rate in gallons per minute?

Solution:

$$\text{Vol, gal} = (8 \text{ ft})(16 \text{ ft})(0.5 \text{ ft})(7.48) = 478.72 \text{ ga}$$
$$\text{gpm} = \frac{478.72 \text{ gal}}{0.5 \text{ min}} = 957.44 \text{ gal/min}$$

e. Calculate the filter backwash rate in gallons per minute per square foot.

Solution:

$$\text{Area} = (8 \text{ ft})(16 \text{ ft}) = 128 \text{ ft}^2$$

$$\text{gpm/ft}^2 = \frac{957.44 \text{ gpm}}{128 \text{ ft}^2} = 7.48 \text{ gpm/ft}^2$$

f. Calculate the gallons of water used to backwash the filter if it was backwashed for 15 min.

Solution:

$$\text{Backwash vol} = (\text{rate, gpm/ft}^2)(\text{time, min})(\text{filter area})$$
$$= (7.48 \text{ gpm/ft}^2)(15 \text{ min})(128 \text{ ft}^2) = 14{,}361.6 \text{ gal}$$

g. During a filter run of 70 hr, the total volume of water filtered was 3.1 million gallons. Calculate the percent of the product water used for backwashing.

Solution:

$$\% \text{ backwash} = \frac{\text{Backwash water}}{\text{Filtered water}} \times 100$$

$$= \frac{14{,}361.6 \text{ gal}}{3100000 \text{ gal}} \times 100$$

$$= 0.46 \%$$

Problem 5.2

A filter basin and its sand bed measures 26 ft by 14 ft. Calculate the sand bed area in square foot.

Solution:

$$\text{Sand bed area, ft}^2 = (\text{length, ft})(\text{width, ft})$$
$$= (26 \text{ ft})(14 \text{ ft})$$
$$= 364 \text{ ft}^2$$

Problem 5.3

The same filter basin that measures 26 ft by 14 ft has the water drop of 6 in. What is the volume in gallons of the drop test?

Solution:

$$6 \text{ in.}/ 12 \text{ in.} = 0.5$$

$$\text{Vol, gal} = (26 \text{ ft})(14 \text{ ft})(0.5 \text{ ft})(7.48 \text{ gal/ft}^3)$$
$$= 1361.4 \text{ gal}$$

Problem 5.4

The filter drop test was timed. The test times were 66 sec, 71 sec, and 70 sec. What was the average time in minutes?

Solution:

$$\text{Avg} = \frac{66 \text{ s} + 71 \text{ s} + 70 \text{ s}}{3} = 69 \text{ seconds}$$

$$= \frac{69 \text{ sec min}}{60 \text{ sec}} = 1.15 \text{ min}$$

Problem 5.5

A filter measures 26 ft by 18 ft. The influent is closed and the effluent is opened, and the water drops 6 in in 2 min. What is the filter rate in gallons per minute?

Solution:

$$6/12 \text{ in.} = 0.5 \text{ ft}$$

$$\text{Vol, gal} = (26 \text{ ft})(18 \text{ ft})(0.5 \text{ ft})(7.48 \text{ gal/ft}^3)$$

$$= \frac{1750.3 \text{ gal}}{2 \text{ min}}$$

$$= 875.16 \text{ gpm}$$

Problem 5.6

A filter measures 26 ft by 18 ft. The influent is closed and the effluent is opened, and the water drains down 6 in in 2 min. What is the filter loading rate in gallons per minute per square foot?

Solution:

$$\text{Vol, gal} = (26 \text{ ft})(18 \text{ ft})(0.5 \text{ ft})(7.48 \text{ gal/ft}^3) = 1750.3 \text{ gal}$$

$$\text{Sand area, ft}^2 = (26 \text{ ft})(18 \text{ ft}) = 468 \text{ ft}^2$$

$$\text{gpm} = \frac{1750.3 \text{ gal}}{2 \text{ minutes}} = 875 \text{ gpm}$$

$$\text{gpm//ft}^2 = \frac{875 \text{ gpm}}{468 \text{ ft}^2}$$

$$= 1.87 \text{ gpm/ft}^2$$

Problem 5.7

A filter measures 28 ft by 14 ft. The influent line is shut, and the water drops 2.8 in per minute. Calculate the rate of filtration in million gallons per day.

Problem:

$$\frac{2.8 \text{ inches}}{12 \text{ inches}} = 0.233 \text{ ft}$$

$$\text{Vol, gal} = (26 \text{ ft})(20 \text{ ft})(1 \text{ ft})(7.48)$$

$$\frac{683.19 \text{ gal}}{\text{min}} \frac{1440 \text{ min}}{\text{day}} \frac{1 \text{ MG}}{1000000 \text{ gal}} = 0.98 \text{ MGD}$$

Problem 5.8

A filter measures 26 ft by 14 ft and has a filter media depth of 36 in. Assuming a backwash rate of 14 gpm/ft² and 10 min of backwash required, how many gallons of water are required for each backwash?

Solution:

$$\text{Backwash vol, gal} = (\text{backwash rate})(\text{time})(\text{filter area})$$
$$= (14 \text{ gpm/ft}^2)(10 \text{ min})(26 \text{ ft})(14 \text{ ft})$$
$$= 50,960 \text{ gal}$$

Problem 5.9

The filter in problem 5.8 filtered 13.90 MG during the last filter run. Based on the gallons produced and the gallons required to backwash the filter, calculate the percent of the product water used for backwashing.

Solution:

$$\% \text{ backwash} = \frac{50,960 \text{ gal}}{13900000 \text{ gal}} \times 100$$
$$= 0.367\%$$

Problem 5.10

Calculate the filtration rate in gallons per minute per square foot for a filter with a sand area of 28 ft by 21 ft when the applied flow is 2.35 MGD.

Solution:

$$\frac{2.35 \text{ MG}}{\text{day}} \frac{1 \text{ day}}{1440 \text{ min}} \frac{1000000 \text{ gal}}{1 \text{ MG}} = 1632 \text{ gpm}$$

$$\text{gpm/ft}^2 = \frac{1632 \text{ gpm}}{(28 \text{ ft})(21 \text{ ft})} = 2.78 \text{ gpm/ft}$$

Problem 5.11

Determine the filtration rate in gallons per minute per square foot for a filter with a surface of 24 ft by 18 ft. With the influent valve closed, the water above the filter dropped 12 in in 5 min.

Solution:

$$\text{Vol, gal} = (26 \text{ ft})(20 \text{ ft})(1 \text{ ft})(7.48)$$

$$= \frac{3889.6 \text{ gal}}{5 \text{ min}} = 777.9 \text{ gal/min}$$

$$\text{gpm/ft}^2 = \frac{777.9 \text{ gpm}}{(24 \text{ ft})(18 \text{ ft})} = 1.8 \text{ gpm/ft}$$

Problem 5.12

A filter measures 25 ft by 12 ft. The influent line is shut, and the water drops 2.4 in/min. Calculate the rate of filtration in million gallons per day.

Solution:

$$2.4/12 = 0.2 \text{ ft}$$

$$\text{Vol, gal} = (25 \text{ ft})(12 \text{ ft})(0.2 \text{ ft})(7.48) = 448.8 \text{ gal/min}$$

$$\frac{448.8 \text{ gal}}{\text{Min}} \frac{1440 \text{ min}}{\text{day}} \frac{1 \text{ MG}}{1000000 \text{ gal}} = 0.65 \text{ MGD}$$

Problem 5.13

The filter in problem 5.12 has a filter media depth of 36 in. Assuming a backwash rater of 14 gpm/ft² and 6 min of backwash, how many gallons of water is required for each backwash?

Solution:

$$\text{Backwash, vol} = (14 \text{ gpm/ft}^2)(6 \text{ min})(25 \text{ ft})(12 \text{ ft}) = 25,200 \text{ gal}$$

Problem 5.14

A filter plant has six filters, each measuring 22 ft by 16 ft. One filter is out of commission. The other five filters are capable of filtering 600 gpm. How many gallons per minute per square foot will each filter?

Solution:

$$\text{gpm/ft}^2 = \frac{600 \text{ gpm}}{(22 \text{ ft})(16 \text{ ft})}$$

$$= 1.7 \text{ gpm/ft}^2$$

Problem 5.15

A filter is 30 ft by 18 ft. If, when the influent value is closed, the water above the filter drops 3.8 in per minute, what is the rate of filtration in million gallons per day?

Solution:

$$3.8 \text{ inc.}/12 = 0.317 \text{ ft}$$

$$\text{gal/min} = \frac{(30 \text{ ft})(18 \text{ ft})(0.317 \text{ ft})(7.48)}{1 \text{ min}}$$

$$= 1280.4 \text{ gal/min}$$

$$\frac{1280.4 \text{ gal}}{\text{Min}} \frac{1440 \text{ min}}{\text{day}} \frac{1 \text{ MG}}{1000000 \text{ gal}} = 1.84 \text{ MGD}$$

Problem 5.16

Determine the backwash pumping rate in gallons per minute for a filter 32 ft by 20 ft if the desired backwash rate is 16 GPM/ft^2.

Solution:

$$\text{gpm} = \left(16 \text{ gpm/ft}^2\right)\left(32 \text{ ft}\right)\left(20 \text{ ft}\right)$$

$$= 10,240 \text{ gpm}$$

Problem 5.17

Determine the volume of water in gallons required to backwash the filter in the previous problem if the filter is backwashed for 5 min.

Solution:

$$\text{Backwash vol} = \left(16 \text{ gpm/ft}^2\right)\left(5 \text{ min}\right)\left(32 \text{ ft}\right)\left(20 \text{ ft}\right)$$

$$= 51,200 \text{ gal}$$

Problem 5.18

During a filter run, the total volume of water filtered was 14.55 MG. When the filter was backwashed, 72,230 gal of water were used. Calculate the percent of the filtered water used for backwashing.

Solution:

$$\% \text{ backwash} = \frac{72,230 \text{ gal}}{14550000 \text{ gal}} \times 100$$

$$= 5\% \text{ (rounded)}$$

5.20 BIBLIOGRAPHY

AWWA & ASCE, 1990. *Water Treatment Plant Design*, 2nd ed. American Water Works Association & American Society of Civil Engineers. New York: McGraw-Hill.

AWWA & ASCE, 1998. *Water Treatment Plant Design*, 3rd ed. American Water Works Association & American Society of Civil Engineers. New York: McGraw-Hill.

Chase, G.L., 2002. *Solids Notes: Fluidization*. Akron, OH: The University of Akron.

Cleasby, J.L., 1990. Filtration. In: *Water Quality and Treatment*. American Water Works Association. New York: McGraw-Hill.

Cleasby, J.L. & Fan, K.S., 1981. Predicting Fluidization and Expansion of Filter Media. *J. Environ. Eng. Div.* **107**(EE3): 355–471. American Society of Civil Engineers.

Downs, A.J. & Adams, C.J., 1973. *The Chemistry of Chlorine, Bromine, Iodine and Astatine*. Oxford: Pergamon.

Droste, R.L., 1997. *Theory and Practice of Water and Wastewater Treatment*. New York: John Wiley & Sons, Inc.

Fair, G.M., Geyer, J.C. & Okun, D.A., 1968. *Water and Wastewater Engineering, vol. 2: Water Purification and Wastewater Treatment and Disposal*. New York: John Wiley.

Fetter, C.W., 1998. *Handbook of Chlorination*. New York: Litton Educational.

Gregory, R. & Zabel, T.R., 1990. Sedimentation and Flotation. In: F.W. Pontius, Ed., *Water Quality and Treatment, a Handbook of Community Water Supplies*, 4th ed. American Water Works Association. New York: McGraw-Hill.

Gupta, R.S., 1997. *Environmental Engineering and Science: An Introduction*. Rockville, MD: Government Institutes.

Hudson, H.E., 1989. Density Considerations in Sedimentation. *J. Amer. Water Works Assoc.* **64**(6): 382–386.

McGhee, T.J., 1991. *Water Resources and Environmental Engineering*, 6th ed. New York: McGraw-Hill.

Morris, J.C., 1966. The Acid Ionization Constant of HOCl from 5°C to 35°C. *J. Phys. Chem.* **70**(12): 3789.

USEPA, 1999. *Individual Filter Self Assessment: EPA Guidance Manual, Turbidity Provisions*. Washington, DC: Environment Protection Agency, pp. 5–12.

Wen, C.Y. & Yu, Y.H., 1966. Minimum Fluidization Velocity. *AIChE J.* **12**(3): 610–612.

White, G.C., 1972. *Handbook of Chlorination*. New York: Litton Education.

White, G.C., 1978. *Disinfection of Wastewater and Water for Reuse*. New York: Van Nostrand Reinhold.

6 Water Chlorination Calculations

Chlorine is the most commonly used substance for disinfection of water in the United States. The addition of chlorine or chlorine compounds to water is called chlorination. Chlorination is considered to be the single most important process for preventing the spread of waterborne disease.

6.1 CHLORINE DISINFECTION

Chlorine deactivates microorganisms through several mechanisms that can destroy most biological contaminants, including:

- It causes damage to the cell wall.
- It alters the permeability of the cell (the ability to pass water in and out through the cell wall).
- It alters the cell protoplasm.
- It inhibits the enzyme activity of the cell so it is unable to use its food to produce energy.
- It inhibits cell reproduction.

Chlorine is available in a number of different forms: (1) as pure elemental gaseous chlorine (a greenish-yellow gas possessing a pungent and irritating odor that is heavier than air, nonflammable, and nonexplosive) which, when released to the atmosphere, is toxic and corrosive; (2) as solid calcium hypochlorite (in tablets or granules); or (3) as a liquid sodium hypochlorite solution (in various strengths). The strength of one form of chlorine over the others for a given water system depends on the amount of water to be treated, the configuration of the water system, the local availability of the chemicals, and the skill of the operator.

One of the major advantages of using chlorine is the effective residual that it produces. A residual indicates that disinfection is completed and the system has an acceptable bacteriological quality. Maintaining a residual in the distribution system helps prevent regrowth of those microorganisms that were inured but not killed during the initial disinfection stage.

6.2 DETERMINING CHLORINE DOSAGE (FEED RATE)

The units of milligrams per liter (mg/L) and pounds per day (lb/day) are most often used to describe the amount of chlorine added or required. Equation 6.1 can be used to calculate either milligrams per liter or pounds per day chlorine dosage.

DOI: 10.1201/9781003354307-6

Chlorine Feed Rate, lb/day $=$ (Chlorine, mg/L) (flow, MGD) (8.34, lb/gal)   (6.1)

Example 6.1

Problem:

Determine the chlorinator setting (in pounds per day) needed to treat a flow of 4 MGD with a chlorine dose of 5 mg/L.

Solution:

$$\text{Chlorine, lbs/day} = (\text{Chlorine, mg/L}) \ (\text{Flow, MGD}) \ (8.34 \text{ lbs/gal})$$
$$= (5 \text{ mg/L}) \ (4 \text{ MGD}) \ (8.34 \text{ lbs/gal})$$
$$= 167 \text{ lbs/day}$$

Example 6.2

Problem:

A pipeline 12 in in diameter and 1,400 ft long is to be treated with a chlorine dose of 48 mg/L. How many pounds of chlorine will this require?

Solution:

Determine the gallon volume of the pipeline.

$$\text{Volume, gal} = (0.785) \ (D^2) \ (\text{Length, ft}) \ (7.48 \text{ gal/cu ft})$$
$$= (0.785) \ (1 \text{ ft}) \ (1 \text{ ft}) \ (1400 \text{ ft}) \ (7.48 \text{ gal/cu ft})$$
$$= 8221 \text{ gal}$$

Now, calculate the pounds chlorine required.

$$\text{lb Chlorine} = (\text{Chlorine, mg/L}) \ (\text{MG Volume}) \ (8.34 \text{ lb/gal})$$
$$= (48 \text{ mg/L}) \ (0.008221 \text{ MG}) \ (8.34 \text{ lb/gal})$$
$$= 3.3 \text{ lb}$$

Example 6.3

Problem:

A chlorinator setting is 30 lb/24 hr. If the flow being chlorinated is 1.25 MGD, what is the chlorine dosage expressed as milligrams per liter?

Solution:

$$\text{Chlorine, lb/day} = (\text{Chlorine, mg/L}) \ (\text{MGD flow}) \ (8.34 \text{ lb/gal})$$
$$30 \text{ lbs/day} = (x \text{mg/L}) \ (\text{flow, MGD}) \ (8.34 \text{ lb/gal})$$
$$x = \frac{30}{(1.25) \ (8.34)}$$
$$x = 2.9 \text{ mg/L}$$

Example 6.4

Problem:

A flow of 1,600 gpm is to be chlorinated. At a chlorinator setting of 48 lb/24 hr, what would be the chlorine dosage in milligrams per liter?

Solution:

Convert the gallons per minute flow rate to million gallons per day flow rate:

$$(1600 \text{ gpm}) \ (1440 \text{ min/day}) = 2{,}304{,}000 \text{ gpd}$$
$$= 2.304 \text{ MGD}$$

Now, calculate the chlorine dosage in milligrams per liter:

$$\text{Chlorine, lb/day} = (\text{Chlorine, mg/L}) \ (\text{Flow, MGD})$$
$$(x \text{ mg/L}) \ (2.304 \text{ MGD}) \ (8.34 \text{ lbs/gal}) = 48 \text{ lbs/day}$$
$$x = \frac{48}{(2.304) \ (8.34)}$$
$$x = 2.5 \text{ mg/L}$$

6.3 CALCULATING CHLORINE DOSE, DEMAND, AND RESIDUAL

Common terms used in chlorination include the following:

- *Chlorine dose*—the amount of chlorine added to the system. It can be determined by adding the desired residual for the finished water to the chlorine demand of the untreated water. Dosage can be either milligrams per liter (mg/L) or pounds per day (lb/day). The most common is milligrams per liter.

> **Chlorine Dose, mg/L = Chlorine Demand, mg/L + Chlorine, mg/L Residual, mg/L**

- *Chlorine demand*—the amount of chlorine used by iron, manganese, turbidity, algae, and microorganisms in the water. Because the reaction between chlorine and microorganisms is not instantaneous, demand is relative to time. For instance, the demand 5 min after applying chlorine will be less than the demand after 20 min. Demand, like dosage, is expressed in milligrams per liter. The chlorine demand is as follows:

> Chlorine Demand = Chlorine dose – Chlorine Residual

- *Chlorine residual*—the amount of chlorine (determined by testing) remaining after the demand is satisfied. Residual, like demand, is based on time. The longer the time after dosage, the lower the residual will be, until all the demand has been satisfied. Residual, like dosage and demand, is expressed in milligrams per liter. The presence of a *free residual* of at least 0.2–0.4 ppm usually provides a high degree of assurance that the disinfection of the water is complete. *Combined residual* is the result of combining free chlorine with nitrogen compounds. Combined residuals are also called chloramines. *Total chlorine residual* is the mathematical combination of free and combined residuals. Total residual can be determined directly with standard chlorine residual test kits.

The following examples, using equation 6.2, illustrate the calculation of chlorine dose, demand, and residual.

Chlorine dose, mg/L = Chlorine demand, mg/L + Chlorine Residual, mg/L   (6.2)

Example 6.5

Problem:

A water sample is tested and found to have a chlorine demand of 1.7 mg/L. If the desired chlorine residual is 0.9 mg/L, what is the desired chlorine dose in milligrams per liter?

Solution:

$$\text{Chlorine dose, mg/L} = \text{Chlorine demand, mg/L} + \text{Chlorine residual, mg/L}$$
$$= 1.7 \text{ mg/L} + 0.9 \text{ mg/L}$$
$$= 2.6 \text{ mg/L Chlorine dose}$$

Example 6.6

Problem:

The chlorine dosage for water is 2.7 mg/L. If the chlorine residual after 30 min contact time is found to be 0.7 mg/L, what is the chlorine demand expressed in milligrams per liter?

Solution:

$$\text{Chlorine dose, mg/L} = \text{Chlorine demand, mg/L} + \text{Chlorine residual, mg/L}$$
$$2.7 \text{ mg/L} = x \text{ mg/L} + 0.6 \text{ mg/L}$$
$$2.7 \text{ mg/L} - 0.7 \text{ mg/L} = x \text{ mg/L}$$
$$x \text{ Chlorine Demand, mg/L} = 2.0 \text{ mg/L}$$

Example 6.7

Problem:

What should the chlorinator setting be (in pounds per day) to treat a flow of 2.35 MGD if the chlorine demand is 3.2 mg/L and a chlorine residual of 0.9 mg/L is desired?

Solution:

Determine the chlorine dosage in milligrams per liter:

Chlorine dose, mg/L = Chlorine demand, mg/L + Chlorine residual, mg/L

$$= 3.2 \text{ mg/L} + 0.9 \text{ mg/L}$$

$$= 4.1 \text{ mg/L}$$

Calculate the chlorine dosage (feed rate) in pounds per day:

Chlorine, lb/day = (Chlorine, mg/L) (Flow, MGD) (8.34 lb/gal)

$$= (4.1 \text{ mg/L}) \ (2.35 \text{ MGD}) \ (8.34 \text{ lb/gal})$$

$$= 80.4 \text{ lb/day Chlorine}$$

6.4   BREAKPOINT CHLORINATION CALCULATIONS

To produce a free chlorine residual, enough chlorine must be added to the water to produce what is referred to as *breakpoint chlorination* (i.e., the point at which near complete oxidation of nitrogen compounds is reached; any residual beyond breakpoint is mostly free chlorine [see Figure 6.1]). When chlorine is added to natural waters, the chlorine begins combining with and oxidizing the chemicals in the water before it begins disinfecting. Although residual chlorine will be detectable in the water, the chlorine will be in the combined form with a weak disinfecting power. As we see in Figure 6.1, adding more chlorine to the water at this point actually decreases the chlorine residual as the additional chlorine destroys the combined chlorine compounds. At this stage, water may have a strong swimming pool or medicinal taste and odor. To avoid this taste and odor, add still more chlorine to

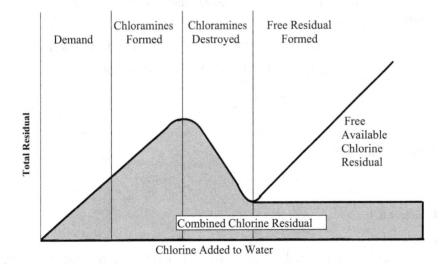

FIGURE 6.1   Breakpoint chlorination curve.

produce a free residual chlorine. Free chlorine has the highest disinfecting power. The point at which most of the combined chlorine compounds have been destroyed and the free chlorine starts to form is the *breakpoint.*

Key Point: The actual chlorine breakpoint of water can only be determined by experimentation.

To calculate the actual increase in chlorine residual that would result from an increase in chlorine dose, we use the milligrams per liter to pounds per day equation, as shown in the following.

$$\text{Increase in Chlorine dose, lb/day} = (\text{Expected Increase, mg/L})$$
$$(\text{Flow, MGD}) \ (8.34 \ \text{lb/gal}) \qquad (6.3)$$

Key Point: The actual increase in residual is simply a comparison of new and old residual data.

Example 6.8

Problem:

A chlorinator setting is increased by 2 lb/day. The chlorine residual before the increased dosage was 0.2 mg/L. After the increased chlorine dose, the chlorine residual was 0.5 mg/L. The average flow rate being chlorinated is 1.25 MGD. Is the water being chlorinated beyond the breakpoint?

Solution:

Calculate the expected increase in chlorine residual. Use the milligrams per liter to pounds per day equation:

$$\text{Lbs/day Increase} = (\text{mg/L Increase}) \ (\text{Flow, MGD}) \ (8.34 \ \text{lbs/gal})$$
$$2 \ \text{lbs/day} = (x \ \text{mg/L}) \ (1.25 \ \text{MGD}) \ (8.34 \ \text{lbs/gal})$$
$$x = \frac{2}{(1.25) \ (8.34)}$$
$$x = 0.19 \ \text{mg/L}$$

Actual increase in residual:

$$0.5 \ \text{mg/L} - 0.19 \ \text{mg/L} = 0.31 \ \text{mg/L}$$

Example 6.9

Problem:

A chlorinator setting of 18 lb chlorine per 24 hr results in a chlorine residual of 0.3 mg/L. The chlorinator setting is increased to 22 lb/24 hr. The chlorine residual increased to 0.4 mg/L at this new dosage rate. The average flow being treated is 1.4 MGD. On the basis of this data, is the water being chlorinated past the breakpoint?

Solution:

Calculate the expected increase in chlorine residual:

$$\text{Lbs/day Increase} = (\text{mg/L Increase}) \ (\text{Flow, MGD}) \ (8.34 \ \text{lbs/gal})$$

$$4 \ \text{lbs/day} = (x \ \text{mg/L}) \ (1.4 \ \text{MGD}) \ (8.34 \ \text{lbs/gal})$$

$$x = \frac{4}{(1.4 \ \text{MGD}) \ (8.34)}$$

$$x = 0.34 \ \text{mg/L}$$

Actual increase in residual:

$$(0.4 \ \text{mg/L} - 0.3 \ \text{mg/L} = 0.1 \ \text{mg/L}$$

6.5   CALCULATING DRY HYPOCHLORITE FEED RATE

The most commonly used dry hypochlorite, calcium hypochlorite, contains about 65–70% available chlorine, depending on the brand. Because hypochlorites are not 100% pure chorine, more pounds per day must be fed into the system to obtain the same amount of chlorine for disinfection. The equation used to calculate the pounds per day hypochlorite needed can be found by:

$$\text{Hypochlorite, lb/day} = \frac{\text{lb/day Chlorine}}{\dfrac{\% \ \text{Available Chlorine}}{100}} \tag{6.4}$$

Example 6.10

Problem:

A chlorine dosage of 110 lb/day is required to disinfect a flow of 1,550,000 gpd. If the calcium hypochlorite to be used contains 65% available chlorine, how many pounds per day hypochlorite will be required for disinfection?

Solution:

Because only 65% of the hypochlorite is chlorine, more than 110 lb of hypochlorite will be required:

$$\text{Hypochlorite, lb/day} = \frac{\text{lb/day Chlorine}}{\dfrac{\% \ \text{Available Chlorine}}{100}}$$

$$= \frac{110 \ \text{lb/day}}{\dfrac{65}{100}}$$

$$= \frac{110}{0.65}$$

$$= 169 \ \text{lb/day Hypochlorite}$$

Example 6.11

Problem:

A water flow of 900,000 gpd requires a chlorine dose of 3.1 mg/L. If calcium hypo-chlorite (65% available chlorine) is to be used, how many pounds per day of hypo-chlorite are required?

Solution:

Calculate the pounds per day chlorine required:

$$\text{Chlorine, lb/day} = (\text{Chlorine, mg/L}) \ (\text{Flow, MGD}) \ (8.34 \text{ lb/gal})$$
$$= (3.1 \text{ mg/L}) \ (0.90 \text{ MGD}) \ (8.34 \text{ lb/gal})$$
$$= 23 \text{ lb/day}$$

The pounds per day hypochlorite:

$$\text{Hypochlorite, lb/day} = \frac{\text{Chlorine, lb/day}}{\dfrac{\% \text{ Available Chlorine}}{100}}$$
$$= \frac{23 \text{ lb/day Chlorine}}{0.65 \text{ Available Chlorine}}$$
$$= 35 \text{ lb/Available day Hypochlorite}$$

Example 6.12

Problem:

A tank contains 550,000 gal of water and is to receive a chlorine dose of 2.0 mg/L. How many pounds of calcium hypochlorite (65% available chlorine) will be required?

Solution:

$$\text{Hypochlorite, lb} = \frac{(\text{mg/L Chlorine}) \ (\text{MG Volume}) \ (8.34 \text{ lb/gal})}{\dfrac{\% \text{ Available Chlorine}}{100}}$$
$$= \frac{(2.0 \text{ mg/L}) \ (0.550 \text{ MG}) \ (8.34 \text{ lb/gal})}{\dfrac{65}{100}}$$
$$= \frac{9.2 \text{ lb}}{0.65}$$
$$= 14.1 \text{ lb Hypochlorite}$$

Example 6.13

Problem:

A total of 40 lb of calcium hypochlorite (65% available chlorine) is used in a day. If the flow rate treated is 1,100,000 gpd, what is the chlorine dosage in milligrams per liter?

Solution:

Calculate the pounds per day chlorine dosage:

$$\text{Hypochlorite, lb/day} = \frac{\text{Chlorine, lb/day}}{\dfrac{\% \text{ Available Chlorine}}{100}}$$

$$40 \text{ lb/day Hypochlorite} = \frac{x \text{ lb/day Chlorine}}{0.65}$$

$$(0.65)\,(40) = x$$

$$26 \text{ lb/day Chlorine} = x$$

Then calculate milligrams per liter chlorine, using the milligrams per liter to pounds per day equation and filling in the known information:

$$26 \text{ lb/day Chlorine} = (x \text{ mg/L Chlorine})\,(1.10 \text{ MGD})\,(8.34 \text{ lb/gal})$$

$$x = \frac{26 \text{ lb/day}}{(1.10 \text{ MGD})\,(8.34 \text{ lb/gal}}$$

$$= 2.8 \text{ mg/L Chlorine}$$

Example 6.14

Problem:

A flow of 2,550,000 gpd is disinfected with calcium hypochlorite (65% available chlorine). If 50 lb of hypochlorite are used in a 24 hr period, what is the milligrams per liter chlorine dosage?

Solution:

Pounds per day chlorine dosage:

$$50 \text{ lb/day Hypochlorite} = \frac{x \text{ lb/day Chlorine}}{0.65}$$

$$x = 32.5 \text{ Chlorine}$$

Calculate the milligrams per liter chlorine:

$$(x \text{ mg/L Chlorine})\,(2.55 \text{ MGD})\,(8.34 \text{ lb/gal}) = 32.5 \text{ lb/day}$$

$$x = 1.5 \text{ mg/L Chlorine}$$

6.6 CALCULATING HYPOCHLORITE SOLUTION FEED RATE

Liquid hypochlorite (i.e., sodium hypochlorite) is supplied as a clear, greenish-yellow liquid in strengths from 5.25% to 16% available chlorine. Often referred to as "bleach," it is, in fact, used for bleaching—common household bleach is a solution of sodium hypochlorite containing 5.25% available chlorine. When calculating gallons per day (gpd) liquid hypochlorite, the pounds per day hypochlorite required must be converted to gallons per day hypochlorite required. This conversion is accomplished using equation 6.5.

$$\text{Hypochlorite, gpd} = \frac{\text{Hypochlorite, lb/day}}{8.34 \text{ lb/gal}} \qquad (6.5)$$

Example 6.15

Problem:

A total of 50 lb/day sodium hypochlorite is required for disinfection of a 1.5 MGD flow. How many gallons per day hypochlorite is this?

Solution:

Because pounds per day hypochlorite has already been calculated, we simply convert pounds per day to gallons per day hypochlorite required:

$$\text{Hypochlorite, gpd} = \frac{\text{Hypochlorite, lb/day}}{8.34 \text{ lb/gal}}$$

$$= \frac{50 \text{ lb/day}}{8.34 \text{ lb/gal}}$$

$$= 6.0 \text{ gpd Hypochlorite}$$

Example 6.16

Problem:

A hypochlorinator is used to disinfect the water pumped from a well. The hypochlorite solution contains 3% available chlorine. A chlorine dose of 1.3 mg/L is required for adequate disinfection throughout the system. If the flow being treated is 0.5 MGD, how many gallons per day of the hypochlorite solution will be required?

Solution:

Calculate the pounds per day chlorine required:

$$(1.3 \text{ mg/L}) \ (0.5 \text{ MGD}) \ (8.34 \text{ lb/gal}) = 5.4 \text{ lb/day Chlorine}$$

Calculate the pounds per day hypochlorite solution required:

$$\text{Hypochlorite, lb/day} = \frac{5.4 \text{ lb/day Chlorine}}{0.03}$$

$$= 180 \text{ lb/day Hypochlorite}$$

Calculate the gallons per day hypochlorite solution required:

$$= \frac{180 \text{ lb/day}}{8.34 \text{ lb/gal}}$$

$$= 21.6 \text{ gpd Hypochlorite}$$

6.7 CALCULATING PERCENT STRENGTH OF SOLUTIONS

If a teaspoon of salt is dropped into a glass of water, it gradually disappears. The salt dissolves in the water. A microscopic examination of the water would not show the salt. Only examination at the molecular level, which is not easily done, would show salt and water molecules intimately mixed. If we taste the liquid, of course, we would know that the salt is there. And we could recover the salt by evaporating the water. In a solution, the molecules of the salt, the *solute,* are homogeneously dispersed among the molecules of water, the *solvent.* This mixture of salt and water is homogenous on a molecular level. Such a homogenous mixture is called a *solution.* The composition of a solution can be varied within certain limits. There are three common states of matter—gas, liquid, and solids. In this discussion, of course, we are only concerned, at the moment, with solids (calcium hypochlorite) and liquid (sodium hypochlorite).

6.7.1 CALCULATING PERCENT STRENGTH USING DRY HYPOCHLORITE

In calculating the percent strength of a chlorine solution, we use equation 6.6.

$$\% \text{ Chlorine Strength} = \frac{\dfrac{(\text{Hypochlorite, lb}) \ (\% \text{ Available Chlorine})}{100}}{\dfrac{\text{Water, lb} + (\text{Hypochlorite, lb}) \ (\% \text{ Available Chlorine})}{100}} \times 100 \qquad (6.6)$$

Example 6.17

Problem:

If a total of 72 oz of calcium hypochlorite (65% available chlorine) is added to 15 gal of water, what is the percent chlorine strength (by weight) of the solution?

Solution:

Convert the ounces of hypochlorite to pounds hypochlorite:

$$\frac{72 \text{ ounces}}{16 \text{ ounces/pound}} = 4.5 \text{ lb chemical}$$

$$\% \text{ Chlorine Strength} = \frac{\dfrac{(\text{Hypochlorite, lb}) \ (\% \text{ Available Chlorine})}{100}}{\dfrac{\text{Water, lb} + (\text{Hypochlorite, lb}) \ (\% \text{ Available Chlorine})}{100}} \times 100$$

$$= \frac{(4.5 \text{ lb}) \ (0.65)}{(15 \text{ gal}) \ (8.34 \text{ lb/gal}) + (4 \text{ lb}) \ (0.65)} \times 100$$

$$= \frac{2.9 \text{ lb}}{125.1 \text{ lb} + 2.6 \text{ lb}} \times 100$$

$$= \frac{(2.6) \ (100)}{127.7}$$

$$= 2.3\% \text{ Chlorine Strength}$$

6.7.2 CALCULATING PERCENT STRENGTH USING LIQUID HYPOCHLORITE

To calculate percent strength using liquid solutions, such as liquid hypochlorite, a different equation is required:

$$\text{Liquid Hypochlorite (gal)} \times 8.34 \text{ lb/gal} \times \frac{\% \text{ Strength of hypochlorite}}{100} = \qquad (6.7)$$

$$\text{Hypochlorite Solution (gal)} \times (8.34 \text{ lb/gal}) \times \frac{\% \text{ Strength of hypochlorite}}{100} =$$

Example 6.18

Problem:

A 12% liquid hypochlorite solution is to be used in making up a hypochlorite solution. If 3.3 gal of liquid hypochlorite are mixed with water to produce 25 gal of hypochlorite solution, what is the percent strength of the solution?

Solution:

Referring to equation 6.7:

$$(3.3 \text{ gal}) \ (8.34 \text{ lb/gal}) \frac{12}{100} = (25 \text{ gal}) \ (8.34 \text{ lb/gal}) \frac{(x)}{100}$$

$$x = \frac{(\cancel{100}) \ (3.3) \ (\cancel{8.34}) \ (12)}{(25) \ (\cancel{8.34}) \ (\cancel{100})}$$

$$x = 1.6\%$$

6.8 CHEMICAL USE CALCULATIONS

In typical waterworks operation, chemical use, in pounds per day or gallons per day, is recorded each day. This data provides a record of daily use from which the average daily use of the chemical or solution can be calculated. To calculate average use in pounds per day, we use equation 6.8. To calculate average use in gallons per day (gpd), we use equation 6.9.

$$\text{Average Use, lb/day} = \frac{\text{Total Chemical Used, lb}}{\text{Number of Days}} \qquad (6.8)$$

$$\text{Average Use, gpd} = \frac{\text{Total Chemicals Used, gal}}{\text{Number of Days}} \qquad (6.9)$$

To calculate the day's supply in inventory, we use equation 6.10 or equation 6.11.

$$\text{Day s Supply in Inventory} = \frac{\text{Total Chemical in Inventory, lb}}{\text{Average Use, lb/day}} \qquad (6.10)$$

$$\text{Day s Supply in Inventory} = \frac{\text{Total Chemical in Inventory, gal}}{\text{Average Use, gpd}} \qquad (6.11)$$

Example 6.19

Problem:

The pounds (lb) calcium hypochlorite used for each day during a week is given in the following. Based on this data, what was the average pounds per day hypochlorite chemical used during the week?

Monday: 50 lb/day	Friday: 56 lb/day
Tuesday: 55 lb/day	Saturday: 51lb/day
Wednesday: 51 lb/day	Sunday: 48 lb/day
Thursday: 46 lb/day	

Solution:

$$\text{Average Use, lb/day} = \frac{\text{Total Chemical Used, lb}}{\text{Number of Days}}$$

$$= \frac{357}{7}$$

$$= 51 \text{ lb/day Average Use}$$

Example 6.20

Problem:

The average calcium hypochlorite use at a plant is 40 lb/day. If the chemical inventory in stock is 1,100 lb, how many days' supply is this?

$$\text{Days' Supply in Inventory} = \frac{\text{Total Chemical in Inventory, lb}}{\text{Average Use, lb/day}}$$

$$\text{Days' Supply in Inventory} = \frac{1100 \text{ lb in Inventory}}{40 \text{ lb/day Average Use}}$$

$$= 27.5 \text{ days' Supply in Inventory}$$

6.9 CHLORINATION MATH PRACTICE PROBLEMS

Problem 6.1

A waterworks wants to have 1.5 mg/L residual chlorine in the distribution system. Due to a main break, the demand has climbed to 0.9 mg/L. What is the required dose?

Solution:

$$\text{Dose} = \text{demand} + \text{residual}$$

$$= 1.5 \text{ mg/L} + 0.9 \text{ mg/L}$$

$$= 2.4 \text{ mg/L}$$

Problem 6.2

A city has a combined residual of 0.6 mg/L and a free residual of 1.6 mg/L. What is the total residual in milligrams per liter?

Solution:

$$Total = combined + residual$$
$$= 0.6 \text{ mg/L} + 1.6 \text{ mg/L}$$
$$= 2.2 \text{ mg/L}$$

Problem 6.3

A water plant treats 4.1 MGD. If the chlorine dose needs to be 4.4 mg/L, what is the chlorine feed requirements in pounds per day?

Solution:

$$lb/day = (dose, \text{ mg/L})(flow, \text{ MGD})(8.34 \text{ lb/gal})$$
$$= (4.4 \text{ mg/L})(4.1 \text{ MGD})(8.34 \text{ lb/gal})$$
$$= 150.45 \text{ lb}$$

Problem 6.4

Determine the chlorine dose in milligrams per liter if 16 lb of chlorine are fed while treating 1.2 MGD of water.

Solution:

$$Dose, \text{ mg/L} = \frac{feed\ rate, \text{ lb/day}}{(flow, \text{ MGD})(8.34 \text{ lb/gal})}$$
$$= \frac{16 \text{ lb}}{(1.2 \text{ MGD})(8.34 \text{ lb/gal})}$$
$$= 1.6 \text{ mg/L}$$

Problem 6.5

How many pounds of 65% available HTH is needed to make 5 gal of an 8% solution?

Solution:

$$HTH, \text{ lbs} = \frac{(Cl)(vol)(8.34 \text{ lbs/gal})}{\% \text{ HTH}}$$
$$= \frac{(0.08)(5 \text{ gal})(8.34 \text{ lb/gal}}{0.65}$$
$$= 5.13 \text{ lb}$$

Problem 6.6

How many gallons of bleach (12.5% available chlorine) will it take to make a 5% solution when added to enough water to make 60 gal of hypochlorite?

Solution:

$$\text{bleach, gpd} = \frac{(\text{Cl})(\text{vol})}{\% \text{ bleach}}$$

$$\frac{(0.05)(60 \text{ gal})}{0.125} = 24 \text{ gal}$$

Problem 6.7

A waterworks has just switched from sodium hypochlorite (bleach) to chlorine gas. If they used an average of 30 gal/day of 15% sodium hypochlorite, how many pounds per day will they use of Cl_2?

Solution:

$$\begin{aligned} \text{Chlorine} &= (\text{avail Cl})(\text{vol})(8.34 \text{ lb/gal}) \\ &= (0.15)(30 \text{ gal})(8.34 \text{ lb/gal}) \\ &= 37.53 \text{ lb} \end{aligned}$$

Problem 6.8

A water system has a chlorine demand of 4.0 mg/L and wants to have a 1.2 mg/L residual. What would be the dose?

Solution:

$$\text{Demand} = \text{dose} - \text{residual}$$

$$4.0 \text{ mg/L} = \text{dose} - 1.2 \text{ mg/L}$$

$$4.0 \text{ mg/L} + 1.2 \text{ mg/L} = \text{dose}$$

$$5.2 \text{ mg/L} = \text{dose}$$

Problem 6.9

A city wants to have 1.5 mg/L chlorine in the distribution system. Due to a main break, the demand has climbed to 1.2 mg/L. What is the residual?

Solution:

$$\text{Demand} = \text{dose} - \text{residual}$$

$$1.2 \text{ mg/L} = 1.5 \text{ mg/L} - \text{residual}$$

$$1.5 \text{ mg/L} - 1.2 \text{ mg/L} = \text{residual}$$

$$= 0.3 \text{ mg/L}$$

Problem 6.10

A system just had a main break. The chlorine level of 3.5 mg/L has dropped to 0.3 mg/L. What is the chlorine demand?

Solution:

$$\text{Demand} = 3.5 \text{ mg/L} - 0.3 \text{ mg/L}$$
$$= 3.2 \text{ mg/L}$$

Problem 6.11

A city doses the water to have a residual of 1.7 mg/L. The demand has risen because of a main break to 1.6 mg/L. What is the free residual?

Solution:

$$\text{Residual} = \text{dose} - \text{demand}$$
$$= 1.7 \text{ mg/L} - 1.6 \text{ mg/L}$$
$$= 0.1 \text{ mg/L}$$

7 Fluoridation

✓ **Note:** The key terms used in this chapter are defined as follows:

- *Fluoride*—found in many waters. It is also added to many water systems to reduce tooth decay.
- *Dental carries*—tooth decay.
- *Dental fluorosis*—result of excessive fluoride content in drinking water causing mottled, discolored teeth.

7.1 WATER FLUORIDATION

As of 1989, fluoridation in the United States has been practiced in approximately 8,000+ communities serving more than 126 million people. Residents of over 1,800 additional communities, serving more than 9 million people, are consuming water that contains at least 0.7 mg/L fluoride from natural sources. Key facts about fluoride include:

- Briefly, fluoride is seldom found in appreciable quantities in surface waters and appears in groundwater in only a few geographical regions.
- Fluoride is sometimes found in a few types of igneous or sedimentary rocks.
- Fluoride is toxic to humans in large quantities; it is also toxic to some animals.
- Based on human experience, fluoride, used in small concentrations (about 1.0 mg/L in drinking water), can be beneficial.

7.2 FLUORIDE COMPOUNDS

Theoretically, any compound that forms fluoride ions in water solution can be used for adjusting the fluoride content of a water supply. However, there are several practical considerations involved in selecting compounds:

- The compound must have sufficient solubility to permit its use in routine water plant practice.
- The cation to which the fluoride ion is attached must not have any undesirable characteristics.
- The material should be relatively inexpensive and readily available in grades of size and purity suitable for their intended use.

Caution: Fluoride chemicals, like chlorine, caustic soda, and many other chemicals used in water treatment can constitute a safety hazard for the water plant operator unless proper handling precautions are observed. It is essential that the operator be aware of the hazards associated with each individual chemical prior to its use.

DOI: 10.1201/9781003354307-7

The three commonly used fluoride chemicals should meet the American Water Works Association (AWWA) standards for use in water fluoridation—sodium fluoride (B701–90), fluorosilicic acid (B703–90), and sodium fluorosilicate (B702–90).

7.2.1 Sodium Fluoride

The first fluoride compound used in water fluoridation was *sodium fluoride*. It was selected based on the previous criteria and also because its toxicity and physiological effects had been so thoroughly studied. It has become the reference standard used in measuring fluoride concentration. Other compounds came into use, but sodium fluoride is still widely used because of its unique physical characteristics. Sodium fluoride (NaF) is a white, odorless material available either as a powder or in the form of crystals of various sizes. It is a salt that, in the past, was manufactured by adding sulfuric acid to fluorspar and then neutralizing the mixture with sodium carbonate. Neutralizing fluorosilicic acid with caustic soda (NaOH) now produces it. Approximately 19 lb of sodium fluoride will add 1 ppm of fluoride to 1 MG of water. Sodium fluoride's solubility is practically constant at 4.0 g/100 mL in water at temperatures generally encountered in water treatment practice (see Table 7.1).

7.2.2 Fluorosilicic Acid

Fluorosilicic acid (H_2SiF_6), also known as hydrofluorosilicic or silicofluoric acid, is a 20–35% aqueous solution with a formula weight of 144.08. It is a straw-colored,

TABLE 7.1
Solubility of Fluoride Chemicals

Chemical	Temperature	Solubility (G per 100 ML of H_2O)
Sodium fluoride	0.0	4.00
	15.0	4.03
	20.0	4.05
	25.0	4.10
	100.0	5.00
Sodium fluorosilicate	0.0	0.44
	25.0	0.76
	37.8	0.98
	65.6	1.52
	100.0	2.45
Fluorosilicic acid	Infinite at all temperatures	

Source: *Water Fluoridation: A Manual for Water Plant Operators.* US Department of Health and Human Services, p. 17, 1994.

TABLE 7.2
Properties of Fluorosilicic Acid

Acid (%)*	Specific Gravity (SG)	Density (Lb/Gal)
0 (water)	1.000	8.345
10	1.0831	9.041
20	1.167	9.739
23	1.191	9.938
25	1.208	10.080
30	1.250	10.431
35	1.291	10.773

* Based on the other percentage being distilled water.

Note: Actual densities and specific gravities will be slightly higher when distilled water is not used.
Add approximately 0.2 lb/gal to density, depending on impurities.

Source: *Water Fluoridation: A Manual for Water Plant Operator.* Washington, DC: US Department of Health and Human Services, p. 19, 1994.

transparent, fuming, corrosive liquid having a pungent odor and an irritating action on the skin. Solutions of 20–35% fluorosilicic acid exhibit a low pH (1.2) and, at a concentration of 1 ppm, can slightly depress the pH of poorly buffered potable waters. It must be handled with great care because it will cause a "delayed burn" on skin tissue. The specific gravity and density of fluorosilicic acid are given in Table 7.2.

It takes approximately 46 lb (4.4 gal) of 23% acid to add 1 ppm of fluoride to 1 MG of water. Two different processes, resulting in products with differing characteristics, manufacture fluorosilicic acid. The largest production of the acid is a by-product of phosphate fertilizer manufacture. Phosphate rock is ground up and treated with sulfuric acid, forming a gas by-product. Hydrofluoric acid (HF) is an extremely corrosive material. Its presence in fluorosilicic acid, whether from intentional addition, that is, as "fortified" acid, or from normal production processes, demands careful handling.

7.2.3   SODIUM FLUOROSILICATE

Fluorosilicic acid can readily be converted into various salts, and one of these, *sodium fluorosilicate* (Na_2SiF_6), also known as sodium silicofluoride, is widely used as a chemical for water fluoridation. As with most fluorosilicates, it is generally obtained as a by-product of the manufacture of phosphate fertilizers. Phosphate rock is ground up and treated with sulfuric acid, thus forming a gas by-product. This gas reacts with water and forms fluorosilicic acid. When neutralized with sodium carbonate, sodium fluorosilicate will precipitate out. The conversion of fluorosilicic acid to a dry material containing a high percentage of available fluoride results in a compound which has most of the advantages of the acid, with few of its disadvantages. Once it was shown that fluorosilicates form fluoride ions in water solution as readily as do simple fluoride compounds, and that there is no difference in the physiological effect, fluorosilicates were rapidly accepted for water fluoridation and, in many cases, have displaced the use of sodium fluoride, except in saturators. Sodium fluorosilicate is a white, odorless, crystalline powder. Its solubility varies (see Table 7.1). Approximately 14 lb of sodium fluorosilicate will add 1 ppm of fluoride to 1 MG of water.

7.3   OPTIMAL FLUORIDE LEVELS

The recommended optimal fluoride concentrations for fluoridated water supply systems are given in Table 7.3. These levels are based on the annual average of the maximum daily air temperature in the area of the involved school or community. In areas where the mean temperature is not shown on Table 7.3, the optimal fluoride level can be determined by the following formula.

$$\text{Fluoride (ppm)} = \frac{0.34}{E} \qquad (7.1)$$

Where:

E = the estimated average daily water consumption for children through 10 years of age in ounces of water per pound of body weight. E is obtained from the formula:

$$E = 0.038 + 0.0062 \text{ average maximum daily air temp. (°F)} \qquad (7.2)$$

TABLE 7.3
Recommended Optimal Fluoride Level

Annual Ave. of Max. Daily Air Temps[1] (°F)	Recommended Fluoride Concentrations		Recommended Control Range			
	Community (ppm)	School[2] (ppm)	Community Systems		School Systems	
			0.1 Below	0.5 Above	20% Low	20% High
40.0–53.7	1.2	5.4	1.1	1.7	4.3	6.5
53.8–58.3	1.1	5.0	1.0	1.6	4.0	6.0
58.4–63.8	1.0	4.5	0.9	1.5	3.6	5.4
63.9–70.6	0.9	4.1	0.8	1.4	3.3	4.9
70.7–79.2	0.8	3.6	0.7	1.3	2.9	4.3
79.3–90.5	0.7	3.2	0.6	1.2	2.6	3.8

[1] Based on temperature data obtained for a minimum of five years.
[2] Based on 4.5 times the optimal fluoride level for communities.
Source: *Water Fluoridation: A Manual for Water Plant Operators.* Washington, DC: US Department of Health and Human Services, p. 21, 1994.

In Table 7.3, the recommended control range is shifted to the high side of the optimal fluoride level for two reasons:

1. It has become obvious that many water plant operators try to maintain the fluoride level in their community at the lowest level possible. The result is that the actual fluoride level in the water will vary around the lowest value in the range instead of around optimal level.
2. Some studies have shown that suboptimal fluorides are relatively ineffective in actually preventing dental caries. Even a drop of 0.2 ppm below optimal levels can reduce dental benefits significantly.

Important Point: In water fluoridation, underfeeding is a much more serious problem than overfeeding.

7.4 FLUORIDATION PROCESS CALCULATIONS

7.4.1 PERCENT FLUORIDE ION IN A COMPOUND

When calculating the percent fluoride ion present in a compound, we need to know the chemical formula for the compound (e.g., NaF) and the atomic weight of each element in the compound. The first step is to calculate the molecular weight of each element in the compound (number of atoms × atomic weight = molecular weight). Then, we calculate the percent fluoride in the compound using equation 7.3.

$$\% \text{ Fluoride in Compound} = \frac{\text{Molecular Weight of Fluoride}}{\text{Molecular Weight of Compound}} \times 100 \quad (7.3)$$

Important Point: *Available fluoride ion* concentration is abbreviated as AFI in the calculations that follow. Other important chemical parameters are listed in Table 7.4.

Example 7.1

Problem:
Given the following data, calculate the percent fluoride in sodium fluoride (NaF).

TABLE 7.4
Fluoride Chemical Parameters

Chemical	Formula	Available Fluoride Ion (AFI) Concentration	Chemical Purity
Sodium fluoride	NaF	0.453	98%
Sodium fluorosilicate	Na_2SiF_6	0.607	98%
Fluorosilicic acid	H_2SiF_6	0.792	23%

Given:

Element	No. of Atoms	Atomic Weight	Molecular Weight
Na	1	22.997	22.997
F	1	19.00	19.00
		Molecular weight of NaF	41.997

Solution:

Calculate the percent fluoride in NaF.

$$\% \text{ F in NaF} = \frac{\text{Molecular Weight of F}}{\text{Molecular Weight of NaF}} \times 100$$

$$= \frac{19.00}{41.997} \times 100$$

$$= 45.2\%$$

Key Point: The molecular weight of hydrofluosilicic acid (H_2SiF_6) is 79.1% and sodium silicofluoride (Na_2SiF_6).

7.4.2 FLUORIDE FEED RATE

Adjusting the fluoride level in a water supply to an optimal level is accomplished by adding the proper concentration of a fluoride chemical at a consistent rate. To calculate the fluoride feed rate for any fluoridation feeder in terms of pounds of fluoride to be fed per day, it is necessary to determine:

- Dosage
- Maximum pumping rate (capacity)
- Chemical purity
- Available fluoride ion concentration

The fluoride feed rate formula is a general equation used to calculate the concentration of a chemical added to water. It will be used for all fluoride chemicals except sodium fluoride when used in a saturator.

Important Point to Remember: Milligrams per liter is equal to parts per million.

The fluoride feed rate (the amount of chemical required to raise the fluoride content to the optimal level) can be calculated as follows:

Fluoride Feed. Rate $(\text{lb/day}) =$

$$\frac{\text{Dosage } (\text{mg/l}) \times \text{cap. } (\text{MGD}) \times 8.34 \text{ lb/gal}}{\text{Available Fluoride Ion } (\text{AFI}) \times \text{chemical purity } (\text{decimal})} \qquad (7.4)$$

If the capacity is in million gallons per day, the fluoride feed rate will be in pounds per day. If the capacity is in gallons per minute, the feed rate will be pounds per minute if a factor of 1 million is included in the denominator.

$$\text{Fluoride Feed Rate } (\text{lb/min}) = \frac{\text{Dosage } (\text{mg/l}) \times \text{cap. } (\text{gpm}) \times 8.34 \text{ lbs/gal}}{1,000,000 \times \text{AFI} \times \text{chemical purity}} \quad (7.5)$$

Example 7.2 (Sodium Fluorosilicate)

Problem:

A water plant produces 2,000 gpm, and the city wants to add 1.1 mg/L of fluoride. What would the fluoride feed rate be?

Solution:

$$2,000 \text{ gpm} \times 1440 \text{ minutes/day} = 2,880,000 \text{ gpd}$$
$$2,880,000 \text{ gpd} \div 1,000,000 = 2.88 \text{ MGD}$$
$$\text{Fluoride Feed Rate } (\text{lb/day}) = \frac{1.1 \text{ mg/l} \times 2.88 \text{ MGD} \times 8.34 \text{ lb/gal}}{0.607 \times 0.985}$$
$$\text{Fluoride Feed Rate, lb/day} = 44.2 \text{ lb/day}$$

The fluoride feed rate is 44.9 lb per day. Some feed rates from equipment design data sheets are given in grams per minute. To convert to grams per minute, divide by 1,440 min/day and multiply by 454 g/lb.

$$\text{fluoride feed rate (gm/min)} = 44.19 \text{ lb/day} \div 1,440 \text{ min/day} \times 454 \text{ gm/lb}$$
$$\text{fluoride feed rate} = 13.9 \text{ gm/min.}$$

Example 7.3 (Fluorosilicic Acid)

Problem:

If it is known that the plant rate is 4,000 gpm and the dosage needed is 0.8 mg/L, what is the fluoride feed rate in milliliters per minute for 23% fluorosilicic acid?

Solution:

$$1,000,000 = 10^6$$

$$\text{Fluoride Fd Rate } (\text{lb/min}) = \frac{\text{Dosage } (\text{mg/l}) \times \text{cap. } (\text{gpm}) \times 8.34 \text{ lb/gal}}{10^6 \times \text{AFI} \times \text{chemical purity}}$$

$$\text{Fluoride Feed Rt } (\text{lb/min}) = \frac{0.8 \text{ mg/l} \times 4,000 \text{ gpm} \times 8.34 \text{ lb/gal}}{10^6 \times 0.79 \times 0.23}$$

$$\text{Fluoride Feed Rate} = 0.147 \text{ lb/min}$$

Note: A gallon of 23% fluorosilicic acid weighs 10 lb, and there are 3,785 mL per gallon; thus, the following formula can be used to convert the feed rate to milliliters per minute:

fluoride feed rate (mL/min) = 0.147 lb/min ÷ 10 lb/gal × 3,785 mL/gal

  fluoride feed rate = 55.6 mL/min

Example 7.4 (Sodium Fluoride)

Problem:

If a small water plant wishes to use sodium fluoride in a dry feeder and the water plant has a capacity (flow) of 180 gpm, what would be the fluoride feed rate?

> **Important Point:** Centers for Disease Control and Prevention (CDC) recommend against using sodium fluoride in a dry feeder.

Assume 0.1 mg/L natural fluoride and 1.0 mg/L is desired in the drinking water.

Solution:

$$\text{Fluoride Feed Rate (lb/min)} = \frac{\text{dosage (mg/l)} \times \text{cap. (gpm)} \times 8.34 \text{ lb/gal}}{10^6 \times \text{AFI} \times \text{chemical purity}}$$

$$\text{Fluoride Feed Rate (lb/min)} = \frac{(1.0 - 0.1) \text{ mg/l} \times 180 \text{ gpm} \times 8.34 \text{ lb/gal}}{10^6 \times 0.45 \times 0.98}$$

$$\text{Fluoride Feed Rate} = 0.003 \text{ lb/min or } 0.18 \text{ lb/hr}$$

Thus, sodium fluoride can be fed at a rate of 0.18 lb/hr to obtain 1.0 mg/L of fluoride in the water.

7.4.3 FLUORIDE FEED RATES FOR SATURATOR

A sodium fluoride saturator is unique in that the strength of the saturated solution formed is always 18,000 ppm. This is because sodium fluoride has a solubility that is practically constant at 4.0 g/100 mL of water at temperatures generally encountered in water treatment. This means that each liter of solution contains 18,000 mg of fluoride ion (40,000 mg/L times the percent available fluoride [45%] equals 18,000 mg/L). This simplifies calculations because it eliminates the need for weighing the chemicals. A meter on the water inlet of the saturator provides this volume; all that is needed is the volume of solution added to the water for calculated dosage.

$$\text{Fluoride Feed Rate (gpm)} = \frac{\text{Cap. (gpm)} \times \text{dosage (mg/l)}}{18,000 \text{ mg/l}} \qquad (7.6)$$

The fluoride feed rate will have the same units as the capacity. If the capacity is in gallons per minute (gpm), the feed rate will be in gallons per minute also. If the capacity is in gallons per day (gpd), the feed rate will be in gallons per day.

Note: For the mathematician, the following derivation is given.

$$\text{Fluoride Feed Rate (lb/min)} = \frac{\text{Dosage (mg/l)} \times \text{cap. (gpm)} \times 8.34 \text{ lb/gal}}{10^6 \times \text{AFI} \times \text{chemical purity}} \quad (7.7)$$

To change the fluoride feed rate from pounds of dry feed to gallons of solution, divide by the concentration of sodium fluoride and the density of the solution (water).

Note: The chemical purity of the sodium fluoride in solution will be 4% × 8.34 lb/gal.

$$\text{Fluoride Feed Rt (gal/min)} = \frac{\text{cap. (gpm)} \times \text{dosage (mg/l)} \times 8.34 \text{ lb/gal}}{10^6 \times \text{AFI} \times \text{chemical purity}}$$

$$\text{Fluoride Feed Rt (gal/min)} = \frac{\text{cap. (gpm)} \times \text{dosage (mg/l)} \times 8.34 \text{ lb/gal}}{10^6 \times 0.45 \times 4\% \times 8.34 \text{ lb/gal}}$$

$$\text{Fluoride Feed Rt (gal/min)} = \frac{\text{cap. (gpm)} \times \text{dosage (mg/l)}}{10^6 \times 0.45 \times 0.04}$$

$$\text{Fluoride Feed Rt (gpm)} = \frac{\text{cap. (gpm)} \times \text{dosage (mg/l)}}{18,000 \text{ mg/12}} \quad (7.8)$$

Example 7.5 (Feed Rate for Saturator)

Problem:

A water plant produces 1.0 MGD and has less than 0.1 mg/L of natural fluoride. What would the fluoride feed rate be to obtain 1.0 mg/L in the water?

Solution:

$$\text{Fluoride Feed Rate (gpd)} = \frac{\text{capacity (gpd)} \times \text{dosage (mg/l)}}{18,000 \text{ mg/l}}$$

$$\text{Fluoride Feed Rate (gpd)} = \frac{1,000,000 \text{ gpd} \times 1.0 \text{ mg/l}}{18,000 \text{ mg/l}}$$

$$\text{Fluoride Feed Rate} = 55.6 \text{ gpd}$$

Thus, it takes approximately 56 gal of saturated solution to treat 1 MG of water at a dose of 1.0 mg/L.

7.4.4 CALCULATED DOSAGES

Some states require that records be kept regarding the amount of chemical used and that the theoretical concentration of chemical in the water be determined mathematically. In order to find the theoretical concentration of fluoride, the calculated dosage must be determined. Adding the calculated dosage to the natural fluoride level in the water supply will yield the theoretical concentration of fluoride in the water. This number, the theoretical concentration, is calculated as a safety precaution to help ensure that an overfeed or accident does not occur. It is also an aid in solving troubleshooting problems. If the theoretical concentration is significantly higher or lower than the measured concentration, steps should be taken to determine the discrepancy. The fluoride feed rate formula can be changed to find the calculated dosage as follows:

$$\text{Dosage (mg/l)} = \frac{\text{Fluoride Fd Rate (lb/day)} \times \text{AFI} \times \text{chemical purity}}{\text{capacity (MGD)} \times 8.34 \text{ lb/gal}} \quad (7.9)$$

When the fluoride feed rate is changed to fluoride fed and the capacity is changed to actual daily production of water in the water system, then the dosage becomes the calculated dosage: the units remain the same, except that fluoride feed goes from pounds per day to pound and actual production goes from million gallons per day to million gallons (MG) (the "day" units cancel).

Note: The amount of fluoride fed (in pounds) will be determined over a time period (day, week, month, etc.), and the actual production will be determined over the same time period.

$$\text{Cal. Dosage (mg/l)} = \frac{\text{Fluoride fed (lb)} \times \text{AFI} \times \text{chemical purity}}{\text{actual production (MG)} \times 8.34 \text{ lb/gal}} \quad (7.10)$$

The numerator of the equation gives the pounds of fluoride ion added to the water, while the denominator gives million pounds of water treated. Pounds of fluoride divided by million pounds of water equals parts per million or milligrams per liter.

The formula for calculated dosage for the saturator is as follows:

$$\text{Calculated Dosage (mg/l)} = \frac{\text{sol. fed (gal)} \times 18,000 \text{ mg/l}}{\text{actual production (gal)}} \quad (7.11)$$

Determining the calculated dosage for an unsaturated sodium fluoride solution is based upon the particular strength of the solution. For example, a 2% strength solution is equal to 8,550 mg/L. The percent strength is based upon the pounds of sodium fluoride dissolved into a certain amount of water. For example, find the percent solution if 6.5 lb of sodium fluoride are dissolved in 45 gal of water:

$$45 \text{ gal} \times 8.34 \text{ lb/gal} = 375 \text{ lb of water}$$

$$\frac{6.5 \text{ lb NaF}}{375 \text{ lb H}_2\text{O}} = 1.7\% \text{ NaF solution}$$

This means that 6.5 lb of fluoride chemical dissolved in 45 gal of water will yield a 1.7% solution. To find the solution concentration of an unknown sodium fluoride solution, use the following formula:

$$\text{Solution Concentration} = \frac{18,000 \text{ mg/l} \times \text{solution strength (\%)}}{4\%} \quad (7.12)$$

Example 7.6

Problem:

Assume that 6.5 lb of NaF are dissolved in 45 gal of water, as previously given. What would be the solution concentration? Solution strength is 1.7% (see earlier example).

Solution:

$$\text{Solution Concentration} = \frac{18,000 \text{ mg/l} \times \text{solution strength } (\%)}{4\%}$$

$$\text{Solution Concentration} = \frac{18,000 \text{ mg/l} \times 1.7\%}{4\%}$$

$$\text{Solution Concentration} = 7,650 \text{ mg/l}$$

Note: The calculated dosage formula for an unsaturated sodium fluoride solution is:

$$\text{Cal.Dosage (mg/l)} = \frac{\text{Sol. fed (gal)} \times \text{sol. conc. (mg/L)}}{\text{Actual production (gal)}}$$

Caution: The CDC recommends against the use of unsaturated sodium fluoride solution in water fluoridation.

7.4.5 CALCULATED DOSAGE PROBLEMS

Example 7.7 (Sodium Fluorosilicate Dosage)

Problem:

A plant uses 65 lb of sodium fluorosilicate in treating 5,540,000 gal of water in one day. What is the calculated dosage?

Solution:

$$\text{Calculated Dosage (mg/l)} = \frac{\text{Fluoride fed (lbs)} \times \text{AFI x purity}}{\text{Actual production (MG)} \times 8.34 \text{ lb/gal}}$$

$$\text{Calculated Dosage (mg/l)} = \frac{65 \text{ lbs} \times 0.607 \times 0.985}{5.54 \text{ MG} \times 8.34 \text{ lbs/gal}}$$

$$\text{Calculated Dosage} = 0.84 \text{ mg/l}$$

Example 7.8 (Fluorosilicic Acid Dosage)

Problem:

A plant uses 43 lb of fluorosilicic acid in treating 1,226,000 gal of water. Assume the acid is 23% purity. What is the calculated dosage?

Solution:

$$\text{Calculated Dosage } (\text{mg/l}) = \frac{\text{Fluoride fed } (\text{lb}) \times \text{AFI x purity}}{\text{Actual production } (\text{MG}) \times 8.34 \text{ lb/gal}}$$

$$\text{Cal.Dosage } (\text{mg/l}) = \frac{43 \text{ lb} \times 0.792 \times 0.23}{1.226 \text{ MG} \times 8.34 \text{ lb/gal}}$$

$$\text{Calculated Dosage} = 0.77 \text{ mg/l}$$

Note: The calculated dosage is 0.77 mg/L. If the natural fluoride level is added to this dosage, then it should equal what the actual fluoride level is in the drinking water.

Example 7.9 (Dry Sodium Fluoride Dosage)

Problem:

A water plant feeds sodium fluoride in a dry feeder. They use 5.5 lb of the chemical to fluoridate 240,000 gal of water. What is the calculated dosage?

Solution:

$$\text{Cal.Dosage } (\text{mg/l}) = \frac{\text{Fluoride fed } (\text{lb}) \times \text{AFI} \times \text{purity}}{\text{Actual production } (\text{MG}) \times 8.34 \text{ lb/gal}}$$

$$\text{Cal.Dosage } (\text{mg/l}) = \frac{5.5 \text{ lbs} \times 0.45 \times 0.98}{0.24 \text{ MG} \times 8.34 \text{ lb/gal}}$$

$$\text{Calculated Dosage} = 1.2 \text{ mg/l}$$

Example 7.10 (Sodium Fluoride Saturator Dosage)

Problem:

A plant uses 10 gal of sodium fluoride from its saturator in treating 200,000 gal of water. What is the calculated dosage?

Solution:

$$\text{Cal.Dosage } (\text{mg/l}) = \frac{\text{Solution fed } (\text{gal}) \times 18,000 \text{ mg/l}}{\text{Actual production } (\text{gal})}$$

$$\text{Cal.Dosage } (\text{mg/l}) = \frac{10 \text{ gallons} \times 18,000 \text{ mg/l}}{200,000 \text{ gallons}}$$

$$\text{Calculated Dosage} = 0.9 \text{ mg/l}$$

Example 7.11 (Sodium Fluoride Unsaturated Solution Dosage)

Problem:

A water plant adds 93 gal per day of a 2% solution of sodium fluoride to fluoridate 800,000 gal/day. What is the calculated dosage?

Solution:

$$\text{Sol. Concentration (mg/l)} = \frac{18,000 \text{ mg/l} \times \text{sol. strength (\%)}}{4\%}$$

$$\text{Sol. Concentration (mg/l)} = \frac{18,000 \text{ mg/l} \times 0.02}{0.04}$$

Solution Concentration = 9,000 mg/l

$$\text{Calculated Dosage (mg/l)} = \frac{\text{solution fed (gal)} \times \text{sol. conc. (mg/l)}}{\text{actual production (gal)}}$$

$$\text{Calculated Dosage (mg/l)} = \frac{93 \text{ gal} \times 9,000 \text{ mg/l}}{800,000 \text{ gal}}$$

Calculated Dosage = 1.05 mg/l

7.4.6 FLUORIDATION MATH PRACTICE PROBLEMS

Problem 7.1

A water plant produces 1,800 gpm, and the town wants to have a 0.9 mg/L of fluoride in the finished water. If fluorosilicic acid is used, what would be the fluoride feed rate in pounds per day?

Note: For the following problems, AFI and purity parameters are found in Table 7.4.

Solution:

$$\frac{1800 \text{ gal}}{\text{min}} \frac{60 \text{ min}}{1 \text{ day}} \frac{1 \text{ MG}}{1000000 \text{ gal}} = 0.108 \text{ MGD}$$

$$\text{Feed rate, lb/day} = \frac{(\text{dose})(\text{flow, MGD})(8.34 \text{ lb/gal})}{(\text{AFI})(\text{purity})}$$

$$= \frac{(0.9 \text{ mg/L})(0.108 \text{ MGD})(8.34 \text{ lb/gal})}{(0.792)(0.23)} = 3.85 \text{ lb/day}$$

Problem 7.2

A water plant produces 2,650 gpm. What would be the fluoride feed rate from a saturator in gallons per minute to obtain 0.7 mg/L in the water?

Solution:

$$\text{Rate, gpm} = \frac{(\text{dose})(\text{flow, gpm})}{18000 \text{ mg/L}}$$

$$= \frac{(0.7 \text{ mg/L})(2650 \text{ gpm})}{18000 \text{ mg/L}} = 0.10 \text{ gpm}$$

Problem 7.3

A plant uses 80 lb of sodium fluorosilicate in treating 9 MGD. What is the calculated dosage in milligrams per liter?

Solution:

$$\text{dose, mg/L} = \frac{(\text{Fl, lbs})(\text{AFI})(\text{Purity})}{(\text{flow, MGD})(8.34 \text{ lb/gal})}$$

$$= \frac{(80 \text{ lb})(0.985)(0.607)}{(9.0 \text{ MGD})(8.34 \text{ lb/gal})} = 0.64 \text{ mg/L}$$

Problem 7.4

The fluoride for a plant's raw water source was measured to be 0.2 mg/L. If the city wants that finished water to contain the recommended amount of 0.6 mg/L, what milligrams per liter of fluoride should the water plant dose?

Solution:

$$\text{Fl} = 0.6 \text{ mg/L} - 0.2 \text{ mg/L} = 0.4 \text{ mg/L}$$

Problem 7.5

A water plant has a daily average production of 660 gpm, and the city wants to have a 1.0 mg/L fluoride in the finished water. The natural fluoride level is less than 0.1 mg/L. Find the fluoride feed rate in pounds per day using sodium fluorosilicate.

Solution:

$$\frac{660 \text{ gal}}{\text{min}} \frac{1440 \text{ min}}{\text{day}} \frac{1 \text{ MG}}{1000000 \text{ gal}} = 0.9504 \text{ MGD}$$

$$\text{lb/day} = \frac{(1.0 \text{ mg/L})(0.9504 \text{ MGD})(8.34 \text{ lbs/gal})}{(0.607)(0.985)} = 13.25 \text{ lb/day}$$

Problem 7.6

If it is known that the plant rate is 4,200 gpm and the dosage needed is 0.9 mg/L, what is the fluoride feed rate in pounds per minute using fluorosilicic acid?

Solution:

$$\text{lb/min} = \frac{(\text{dose})(\text{flow})(8.34 \text{ lb/gal})}{(1000000)(\text{AFI})(\text{purity})}$$

$$= \frac{(0.9 \text{ mg/L})(4200 \text{ gpm})(8.34 \text{ lb/gal})}{(1000000)(0.79)(0.23)} = 0.17 \text{ lb/min}$$

Problem 7.7

What is the fluoride feed rate in pounds per day using fluorosilicic acid if the plant rate is 1 MGD, the natural fluoride content is 0.3 mg/L, and the desired fluoride content is 1.3 mg/L?

Solution:

$$\text{Dose} = 1.3 \text{ mg/L} - 0.3 \text{ mg/L} = 1.0 \text{ mg/L}$$

$$\text{lb/day} = \frac{(1.0 \text{ mg/L})(1.0 \text{ MGD})(8.34 \text{ lb/gal})}{(0.792)(0.23)} = 45.8 \text{ lb/day}$$

Problem 7.8

If a waterworks wishes to use sodium fluorosilicate in a dry feeder and the water plant has a flow of 210 gpm, what would the fluoride feed rate be in pounds per minute? Assume 0.1 mg/L natural fluoride and 1.1 mg/L is the desired concentration in the finished water.

Solution:

$$\text{Dose} = 1.1 \text{ mg/L} - 0.1 \text{ mg/L} = 1.0 \text{ mg/L}$$

$$\text{lb/min} = \frac{(1.0 \text{ mg/L})(210 \text{ gpm})(8.34 \text{ lb/gal})}{(1000000)(0.607)(0.985)} = 0.003 \text{ lb/min}$$

Problem 7.9

A waterworks produces 1.1 MGD. What would the fluoride feed rate be from a saturator in gallons per day to obtain 1.1 mg/L in the water?

Solution:

$$\frac{1.1 \text{ MG}}{\text{Day}} \frac{1 \text{ day}}{1440 \text{ min}} \frac{1000000 \text{ gal}}{1 \text{ MG}} = 764 \text{ gpm}$$

$$\text{gal/min} = \frac{(\text{dose})(\text{flow})}{18000 \text{ mg/L}}$$

$$\text{gal/min} = \frac{(1.1 \text{ mg/L})(764 \text{ gpm})}{18000 \text{ mg/L}} = 0.047 \text{ gpm}$$

$$\frac{0.047 \text{ gal}}{\text{min}} \frac{1440 \text{ min}}{\text{day}} = 67.7 \text{ gpd}$$

8 Water Softening

8.1 WATER HARDNESS

Hardness in water is caused by the presence of certain positively charged metallic ions in solution in the water. The most common of these hardness-causing ions are calcium and magnesium; others include iron, strontium, and barium. The two primary constituents of water that determine the hardness of water are calcium and magnesium. If the concentration of these elements in the water is known, the total hardness of the water can be calculated. To make this calculation, the equivalent weights of calcium, magnesium, and calcium carbonate must be known; the equivalent weights are given in the following.

EQUIVALENT WEIGHTS

Calcium, Ca	**20.04**
Magnesium, Mg	**12.15**
Calcium Carbonate, $CaCO_3$	**50.045**

8.1.1 CALCULATING CALCIUM HARDNESS, AS $CaCO_3$

The hardness (in milligrams per liter as $CaCO_3$) for any given metallic ion is calculated using equation 8.1.

$$\frac{\text{Calcium Hardness, mg/L as } CaCO_3}{\text{Equivalent Weight of } CaCO_3} = \frac{\text{Calcium, mg/L}}{\text{Equivalent Weight of Calcium}} \quad (8.1)$$

Example 8.1

Problem:

A water sample has calcium content of 51 mg/L. What is this calcium hardness expressed as $CaCO_3$?

Solution:

$$\frac{\text{Calcium Hardness, mg/L as } CaCO_3}{\text{Equivalent Weight of } CaCO_3} = \frac{\text{Calcium, mg/L}}{\text{Equivalent Weight of Calcium}}$$

$$\frac{x \, \text{mg/L}}{50.045} = \frac{51 \, \text{mg/L}}{20.04}$$

$$x = \frac{(51) \, (50.045)}{20.45}$$

$$x = 124.8 \text{ mg/L Ca as } CaCO_3$$

DOI: 10.1201/9781003354307-8

Example 8.2

Problem:

The calcium content of a water sample is 26 mg/L. What is this calcium hardness expressed as $CaCO_3$?

Solution:

$$\frac{\text{Calcium Hardness, mg/L as CaCO}_3}{\text{Equivalent Weight of CaCO}_3} = \frac{\text{Calcium, mg/L}}{\text{Equivalent Weight of Calcium}}$$

$$\frac{x\,\text{mg/L}}{50.045} = \frac{26\,\text{mg/L}}{20.04}$$

$$x = \frac{(26)\,(50.045)}{20.04}$$

$$x = 64.9 \text{ mg/L Ca as CaCO}_3$$

8.1.2 CALCULATING MAGNESIUM HARDNESS, AS $CaCO_3$

To calculate magnesium hardness, we use equation 8.2.

$$\frac{\text{Magnesium Hardness, m/L as CaCO}_3}{\text{Equivalent Weight of CaCO}_3} = \frac{\text{Magnesium, mg/L}}{\text{Equivalent Weight of Magnesium}} \quad (8.2)$$

Example 8.3

Problem:

A sample of water contains 24 mg/L magnesium. Express this magnesium hardness as $CaCO_3$.

Solution:

$$\frac{\text{Magnesium Hardness, m/L as CaCO}_3}{\text{Equivalent Weight of CaCO}_3} = \frac{\text{Magnesium, mg/L}}{\text{Equivalent Weight of Magnesium}}$$

$$\frac{x\,\text{mg/L}}{50.045} = \frac{24\,\text{mg/L}}{12.15}$$

$$x = \frac{(24)\,(50.045)}{12.15}$$

$$x = 98.9 \text{ mg/L}$$

Example 8.4

Problem:

The magnesium content of a water sample is 16 mg/L. Express this magnesium hardness as $CaCO_3$.

Solution:

$$\frac{\text{Magnesium Hardness, mg/L as CaCO}_3}{\text{Equivalent Weight of CaCO}_3} = \frac{\text{Magnesium, mg/L}}{\text{Equivalent Weight of Magnesium}}$$

$$\frac{x\,\text{mg/L}}{50.045} = \frac{16\,\text{mg/L}}{12.15}$$

$$x = \frac{(16)\,(50.045)}{12.15}$$

$$x = 65.9 \text{ mg/L Mg as CaCO}_3$$

8.1.3 CALCULATING TOTAL HARDNESS

Calcium and magnesium ions are the two constituents that are the primary cause of hardness in water. To find total hardness, we simply add the concentrations of calcium and magnesium ions, expressed in terms of calcium carbonate, $CaCO_3$, using equation 8.3.

$$\text{Total Hard., mg/L as CaCO}_3 = \text{Cal. Hard., mg/L as CaCO}_3 + \text{Mag. Hard.,}$$
$$\text{mg/L as CaCO}_3 \tag{8.3}$$

Example 8.5

Problem:

A sample of water has calcium content of 70 mg/L as $CaCO_3$ and magnesium content of 90 mg/L as $CaCO_3$.

Solution:

$$\text{Total Hardness, mg/L as CaCO}_3 = \text{Calcium Hard., mg/L} + \text{Magnesium Hard., mg/L}$$
$$= 70 \text{ mg/L} + 90 \text{ mg/L}$$
$$= 160 \text{ mg/L as CaCO}_3$$

Example 8.6

Problem:

Determine the total hardness as $CaCO_3$ of a sample of water that has calcium content of 28 mg/L and magnesium content of 9 mg/L.

Solution:

Express calcium and magnesium in terms of $CaCO_3$:

$$\frac{\text{Calcium Hardness, mg/L as CaCO}_3}{\text{Equivalent Weight of CaCO}_3} = \frac{\text{Calcium, mg/L}}{\text{Equivalent Weight of Calcium}}$$

$$\frac{x\,\text{mg/L}}{50.045} = \frac{28\,\text{mg/L}}{20.04}$$

$$x = 69.9 \text{ mg/L Mg as CaCO}_3$$

$$\frac{\text{Magnesium Hardness, mg/L as CaCO}_3}{\text{Equivalent Weight of CaCO}_3} = \frac{\text{Magnesium, mg/L}}{\text{Equivalent Weight of Magnesium}}$$

$$\frac{x\,\text{mg/L}}{50.045} = \frac{9\,\text{mg/L}}{12.15}$$

$$x = 37.1 \text{ mg/L Mg as CaCO}_3$$

Now, total hardness can be calculated:

$$\text{Total Hardness, mg/L as CaCO}_3 = \text{Cal.Hardness, mg/L} + \text{Mag. Hardness, mg/L}$$
$$= 69.9 \text{ mg/L} + 37.1 \text{ mg/L}$$
$$= 107 \text{ mg/L as CaCO}_3$$

8.1.4 CALCULATING CARBONATE AND NONCARBONATE HARDNESS

As mentioned, total hardness is comprised of calcium and magnesium hardness. Once total hardness has been calculated, it is sometimes used to determine another expression hardness—carbonate and noncarbonate. When hardness is numerically greater than the sum of bicarbonate and carbonate alkalinity, that amount of hardness equivalent to the total alkalinity (both in units of mg $CaCO_3$/L) is called the *carbonate hardness*; the amount of hardness in excess of this is the *noncarbonate hardness*. When the hardness is numerically equal to or less than the sum of carbonate and noncarbonate alkalinity, all hardness is carbonate hardness and noncarbonate hardness is absent.

Again, the total hardness is comprised of carbonate hardness and noncarbonate hardness:

$$\text{Total Hardness} = \text{Carbonate Hardness} + \text{Noncarbonate Hardness} \qquad (8.4)$$

When the alkalinity (as $CaCO_3$) is greater than the total hardness, all the hardness is carbonate hardness:

$$\text{Total Hardness, mg/L as CaCO}_3 = \text{Carbonate Hardness, mg/L as CaCO}_3 \qquad (8.5)$$

When the alkalinity (as $CaCO_3$) is less than the total hardness, then the alkalinity represents carbonate hardness and the balance of the hardness is noncarbonate hardness:

$$\text{Total Hard., mg/L as CaCO}_3 = \text{Carb. Hard., mg/L as CaCO}_3 \\ + \text{Noncarb. Hard., mg/L as CaCO}_3 \qquad (8.6)$$

When carbonate hardness is represented by the alkalinity, we use equation 8.7.

$$\text{Total Hard., mg/L as CaCO}_3 = \text{Alk., mg/L as CaCO}_3 \\ + \text{Noncarb. Hardness, mg/L as CaCO}_3 \qquad (8.7)$$

Example 8.7

Problem:

A water sample contains 110 mg/L alkalinity as $CaCO_3$ and 105 mg/L total hardness as $CaCO_3$. What is the carbonate and noncarbonate hardness of the sample?

Solution:

Because the alkalinity is greater than the total hardness, all the hardness is carbonate hardness:

$$\text{Total Hardness, mg/L as } CaCO_3 = \text{Carbonate Hardness, mg/L as } CaCO_3$$
$$105 \text{ mg/L as } CaCO_3 = \text{Carbonate Hardness}$$

No noncarbonate hardness present in this water.

Example 8.8

Problem:

The alkalinity of a water sample is 80 mg/L as $CaCO_3$. If the total hardness of the water sample is 112 mg/L as $CaCO_3$, what is the carbonate and noncarbonate hardness in milligrams per liter as $CaCO_3$?

Solution:

Alkalinity is less than total hardness. Therefore, both carbonate and noncarbonate hardness will be present in the hardness of the sample.

$$\text{Total Hard., mg/L } CaCO_3 = \text{Carb. Hard., mg/L as } CaCO_3$$
$$+ \text{Noncarb. Hard., mg/L as } CaCO_3$$
$$112 \text{ mg/L} = 80 \text{ mg/L} - x \text{ mg/L}$$
$$112 \text{ mg/L} - 80 \text{ mg/L} = x \text{ mg/L}$$
$$32 \text{ mg/L noncarbonate hardness} = x$$

8.2   ALKALINITY DETERMINATION

Alkalinity measures the acid-neutralizing capacity of a water sample. It is an aggregate property of the water sample and can be interpreted in terms of specific substances only when a complete chemical composition of the sample is also performed. The alkalinity of surface waters is primarily due to the carbonate, bicarbonate, and hydroxide content and is often interpreted in terms of the concentrations of these constituents. The higher the alkalinity, the greater the capacity of the water to neutralize acids; and conversely, the lower the alkalinity, the less the neutralizing capacity. To detect the different types of alkalinity, the water is tested for phenolphthalein and total alkalinity, using equations 8.8 and 8.9.

$$\text{Phenolphthalein Alkalinity mg/L as } CaCO_3 = \frac{(A)\ (N)\ (50,000)}{\text{ML of Sample}} \quad (8.8)$$

$$\text{Total Alkalinity mg/L as } CaCO_3 = \frac{(B)(N)(50,000)}{\text{ML of Sample}} \quad (8.9)$$

Where:
  A = mL titrant used to pH 8.3
  B = total mL of titrant used to titrate to pH 4.5

N = normality of the acid (0.02 N H_2SO_4 for this alkalinity test)

$50,000$ = a conversion factor to change the normality into units of $CaCO_3$

Example 8.9

Problem:

A 100 mL water sample is tested for phenolphthalein alkalinity. If 1.3 mL titrant is used to pH 8.3 and the sulfuric acid solution has a normality of 0.02N, what is the phenolphthalein alkalinity of the water?

Solution:

$$\text{Phenolphthalein Alkalinity, mg/L as } CaCO_3 = \frac{(A)(N)(50,000)}{\text{mL of Sample}}$$

$$= \frac{(1.3 \text{ mL}) \ (0.02N) \ (50,000)}{100 \text{ mL}}$$

$$= 13 \text{ mg/L as } CaCO_3 \text{ Phenolphthalein Alk.}$$

Example 8.10

Problem:

A 100 mL sample of water is tested for alkalinity. The normality of the sulfuric acid used for titrating is 0.02N. If 0 mL is used to pH 8.3 and 7.6 mL titrant is used to pH 4.5, what is the phenolphthalein and total alkalinity of the sample?

Solution:

$$\text{Phenolphthalein Alk. mg/L as } CaCO_3 = \frac{(0 \text{ mL}) \ (0.02N) \ (50,000)}{100 \text{ mL}}$$

$$= 0 \text{ mg/L}$$

$$\text{Total Alkalinity, mg/L as } CaCO_3 = \frac{(7.6 \text{ mL}) \ (0.02N) \ (50,000)}{100 \text{ mL}}$$

$$= 76 \text{ mg/L}$$

8.3 DETERMINING BICARBONATE, CARBONATE, AND HYDROXIDE ALKALINITY

Interpretation of phenolphthalein and total alkalinity test results (assuming all the alkalinity found is due to carbonate, bicarbonate, or hydroxide) can be made using calculations based on the values given in Table 8.1.

Example 8.11

Problem:

A water sample is tested for phenolphthalein and total alkalinity. If the phenolphthalein alkalinity is 10 mg/L as $CaCO_3$ and the total alkalinity is 52 mg/L as $CaCO_3$, what is the bicarbonate, carbonate, and hydroxide alkalinity of the water?

TABLE 8.1
Interpretation of Results Values

Alkalinity, Mg/L as $CaCO_3$

Results of Titration	Bicarbonate Alkalinity	Carbonate Alkalinity	Hydroxide Alkalinity
P = O	T	O	O
P < ½ T	T – 2P	2P	O
P = ½ T	O	2P	O
P > ½ T	O	2T – 2P	2P – T
P = T	O	O	T

Where:

P = phenolphthalein alkalinity

T = total alkalinity

Source: APHA. *Standard Methods*, from Table 2320:II, vol. 19, pp. 2–28, 1995. Washington, DC: American Public Health Association.

Solution:

Based on titration test results, P alkalinity (10 mg/L) is less than half of the T alkalinity (52 mg/L ÷ 2 = 26 mg/L; from Table 8.1). Therefore, each type of alkalinity is calculated as follows:

$$\text{Bicarbonate Alkalinity} = T - 2P$$
$$= 52 \text{ mg/L} - 2(10 \text{ mg/L})$$
$$= 52 \text{ mg/L} - 20 \text{ mg/L}$$
$$= 32 \text{ mg/L as } CaCO_3$$
$$\text{Carbonate Alkalinity} = 2P$$
$$= 2(10 \text{ mg/L})$$
$$= 20 \text{ mg/L as } CaCO_3$$
$$\text{Hydroxide Alkalinity} = 0 \text{ mg/L as } CaCO_3$$

Example 8.12

Problem:

Alkalinity titrations on a water sample resulted as follows:

Sample was 100 mL.

1.4 mL titrant used to pH 8.3.

2.4 mL total titrant used to pH 4.5.

Acid normality was 0.02N H_2SO_4.

What is the phenolphthalein, total bicarbonate, carbonate, and hydroxide alkalinity?

Solution:

$$\text{Phenolphthalein Alkalinity mg/L as CaCO}_3 = \frac{(1.4 \text{ mL}) \ (0.02\text{N}) \ (50{,}000)}{100 \text{ mL}}$$

$$= 14 \text{ mg/L as CaCO}_3$$

$$\text{Total Alkalinity mg/L as CaCO}_3 = \frac{(2.4 \text{ mL}) \ (0.02\text{N}) \ (50{,}000)}{100 \text{ mL}}$$

$$= 24 \text{ mg/L as CaCO}_3$$

Now use Table 8.1 to calculate the other alkalinity constituents (*P* is greater than ½ T).

$$\text{Bicarbonate Alkalinity} = 0 \text{ mg/L as CaCO}_3$$

$$\text{Carbonate Alkalinity} = 2T - 2P$$

$$= 2(24 \text{ mg/L}) - 2(14 \text{ mg/L})$$

$$= 20 \text{ mg/L as CaCO}_3$$

$$\text{Hydroxide Alkalinity} = 2P - T$$

$$= 2(14 \text{ mg/L}) - (24 \text{ mg/L})$$

$$= 4 \text{ mg/L as CaCO}_3$$

8.4 LIME DOSAGE CALCULATION FOR REMOVAL OF CARBONATE HARDNESS

The lime-soda ash water-softening process uses lime, Ca $(OH)_2$, and soda ash, Na_2CO_3, to precipitate hardness from solution. Carbonate hardness (calcium and magnesium bicarbonates) is complexed by lime. Noncarbonate hardness (calcium and magnesium sulfates or chlorides) requires the addition of soda ash for precipitation. The molecular weights of various chemicals and compounds used in lime-soda as softening calculations are as follows:

Quicklime, CaO$_2$	= 56
Hydrated lime, Ca (OH)$_2$	= 74
Magnesium, Mg^{+2}	= 24.3
Carbon dioxide, CO$_2$	= 44
Magnesium hydroxide, Mg (OH)$_2$	= 58.3
Soda ash, Na$_2$CO$_3$	= 100
Alkalinity, as CaCO$_3$	= 100
Hardness, as CaCO$_3$	= 100

To calculate quicklime or hydrated lime dosage, in milligrams per liter, use equation 8.10.

$$\text{Quicklime (CaO) Feed, mg/L} = \frac{(A+B+C+D) \ 1.15}{\dfrac{\% \text{ Purity of Lime}}{100}} \qquad (8.10)$$

Where:

   $A = CO_2$ in source water (mg/L as CO_2) (56/44)
   $B =$ bicarbonate alkalinity removed in softening (mg/L as $CaCO_3$) (56/100)
   $C =$ hydroxide alkalinity in softener effluent (mg/L as $CaCO_3$) (56/100)
   $D =$ magnesium removed in softening (mg/L as Mg^{+2}) (56/24.3)
   $1.15 =$ excess lime dosage (using 15% excess)

Note: For hydrated lime dosage, use equation 8.10 as given for quicklime, except substitute 74 for 56 in A, B, C, and D.

Example 8.13

Problem:

A water sample has a carbon dioxide content of 4 mg/L as CO_2, total alkalinity of 130 mg/L as $CaCO_3$, and magnesium content of 26 mg/L as Mg^{+2}. Approximately how much quicklime (CaO) (90% purity) will be required for softening? (Assume 15% excess lime.)

Solution:

Calculate the A–D factors:

$A = (CO_2, mg/L) (56/44)$ $C = 0$ mg/L
$\quad = (4\ mg/L) (56/44)$
$\quad = 5.1\ mg/L$ $D = (Mg^{+2}, mg/L) (56/24.3)$
$B = (Alkalinity, mg/L) (56/100)$ $\quad = (26\ mg/L) (56/24.3)$
$\quad = (130\ mg/L) (56/100)$ $\quad = 60\ mg/L$
$\quad = 73\ mg/L$

Calculate the estimated quicklime dosage:

$$\text{Quicklime Dosage, mg/L} = \frac{\left(5\ mg/L + 73\ mg/L + 0 + 60\ mg/L\right)\ (1.15)}{0.90}$$

$$= 176\ mg/L\ CaO$$

Example 8.14

Problem:

The characteristics of a water sample are as follows: 4 mg/L CO_2 as CO_2, 175 mg/L total alkalinity as $CaCO_3$, and 20 mg/L magnesium as Mg^{+2}. What is the estimated hydrated lime $(Ca\ (OH)_2)$ (90% pure) dosage in milligrams per liter required for softening? (Assume 15% excess lime.)

Solution:

Determine the A–D factors:
$A = (CO_2, mg/L)(74/44)$ $C = 0$ mg/L
$\quad = (4\ mg/L)(74/44) \ldots$
$\quad = 7\ mg/L$ $D = (Mg^{+2}, mg/L)(74/24.3)$
$B = (Alkalinity, mg/L)(74/100)$ $\quad = (20\ mg/L)(74/24.3)$
$\quad = (175\ mg/L)(74/100)$ $\quad = 61\ mg/L$
$\quad = 130\ mg/L$

Calculate the estimated hydrated lime dosage:

$$\text{Hydrated Lime Dosage, mg/L} = \frac{(7\ \text{mg/L} + 130\ \text{mg/L} + 0 + 61\ \text{mg/L})\ (1.15)}{0.90}$$

$$= 253\ \text{mg/L Ca} \left(\text{OH}\right)_2$$

8.5 CALCULATION FOR REMOVAL OF NONCARBONATE HARDNESS

Soda ash is used for precipitation and removal of noncarbonate hardness. To calculate the soda ash dosage required, we use, in combination, equations 8.11 and 8.12.

$$\text{Tot. Hard., mg/L as CaCO}_3 = \text{Carb. Hard., mg/L as CaCO}_3 \tag{8.11}$$
$$+ \text{Noncarb. Hard., mg/L as CaCO}_3$$

then

$$\text{Soda Ash } (\text{Na}_2\text{CO}_3)\ \text{Fd., mg/L} = \tag{8.12}$$
$$(\text{Noncarb.})\ \text{Hard., mg/L as CaCO}_3\ (106/100)$$

Example 8.15

Problem:

A water sample has a total hardness of 250 mg/L as $CaCO_3$ and a total alkalinity of 180 mg/L. What soda ash dosage (in milligrams per liter) will be required to remove the noncarbonate hardness?

Solution:

Calculate the noncarbonate hardness:
total hardness, mg/L as $CaCO_3$ = carb. hard., mg/L as $CaCO_3$ + noncarb. hard., mg/L as $CaCO_3$

$$250\ \text{mg/L} - 180\ \text{mg/L} = x\ \text{mg/L}$$
$$70\ \text{mg/L} = x$$

Calculate the soda ash required:

$$\text{Soda Ash, mg/L} = (\text{Noncarbonate Hardness}), \text{mg/L as CaCO}_3\ (106)/100$$
$$= (70\ \text{mg/L})\ (106)/100$$
$$= 74.2\ \text{mg/L soda ash}$$

Example 8.16

Problem:

Calculate the soda ash required (in milligrams per liter) to soften water if the water has a total hardness of 192 mg/L and a total alkalinity of 103 mg/L.

Solution:

Determine noncarbonate hardness:

$$192 \text{ mg/L} = 103 \text{ mg/L} + x \text{ mg/L}$$
$$192 \text{ mg/L} - 103 \text{ mg/L} = x$$
$$89 \text{ mg/L} = x$$

Calculate soda ash required:

$$\text{Soda Ash, mg/L} = (\text{Noncarbonate}) \text{ Hardness, mg/L as } CaCO_3 \, (106)/100$$
$$= (89 \text{ mg/L}) \, (106)/100$$
$$= 94 \text{ mg/L soda ash}$$

8.6   RECARBONATION CALCULATION

Recarbonation involves the reintroduction of carbon dioxide into the water, either during or after lime softening, lowering the pH of the water to about 10.4. After the addition of soda ash, recarbonation lowers the pH of the water to about 9.8, promoting better precipitation of calcium carbonate and magnesium hydroxide. Equations 8.13 and 8.14 are used to estimate carbon dioxide dosage.

$$\text{Excess Lime, mg/L} = (A + B + C + D) \, (0.15) \qquad\qquad (8.13)$$

$$\text{Total } CO_2 \text{ Dosage, mg/} = \left[Ca \, (OH)_2 \text{ Excess, mg/L} \right] \, (44)/74$$
$$+ (Mg^{+2}) \text{ Residual mg/L } (44)/24.3 \qquad (8.14)$$

Example 8.17

Problem:

The A, B, C, and D factors of the excess lime equation have been calculated as follows: $A = 14$ mg/L; $B = 126$ mg/L; $C = 0$; $D = 66$ mg/L. If the residual magnesium is 5 mg/L, what is the carbon dioxide (in milligrams per liter) required for recarbonation?

Solution:

Calculate the excess lime concentration:

$$\text{Excess Lime, mg/L} = (A + B + C + D) \, (0.15)$$
$$= (14 \text{ mg/L} + 126 \text{ mg/L} + 0 + 66 \text{ mg/L}) \, (0.15)$$
$$= 31 \text{ mg/L}$$

Determine the required carbon dioxide dosage:

$$\text{Total } CO_2 \text{ Dosage, mg/L} = (31 \text{ mg/L}) \, (44)/74 + (5 \text{ mg/L}) \, (44)/24.3$$
$$= 18 \text{ mg/L} + 9 \text{ mg/L}$$
$$= 27 \text{ mg/L } CO_2$$

Example 8.18

Problem:

The A, B, C, and D factors of the excess lime equation have been calculated as: A = 10 mg/L; B = 87 mg/L; C = 0; D =111 mg/L. If the residual magnesium is 5 mg/L, what carbon dioxide dosage would be required for recarbonation?

Solution:

The excess lime is:

$$\text{Excess Lime, mg/L} = (A+B+C+D)\ (0.15)$$
$$= (10\ \text{mg/L} + 87\ \text{mg/L} + 0 + 111\ \text{mg/L})\ (0.15)$$
$$= (208)\ (0.15)$$
$$= 31\ \text{mg/L}$$

The required carbon dioxide dosage for recarbonation is:

$$\text{Total CO}_2 \text{ Dosage, mg/L} = (31\ \text{mg/L})\ (44)/74 + (5\ \text{mg/L})\ (44)/24.3$$
$$= 18\ \text{mg/L} + 9\ \text{mg/L}$$
$$= 27\ \text{mg/L CO}_2$$

8.7 CALCULATING FEED RATES

The appropriate chemical dosage for various unit processes is typically determined by lab or pilot scale testing (e.g., jar testing, pilot plant), only monitoring, and historical experience. Once the chemical dosage is determined, the feed rate can be calculated by using equation 8.15. Once the chemical feed rate is known, this value must be translated into a chemical feeder setting.

$$\text{Feed Rate (lb/day)} = \text{Flow Rate, MGD} \times \text{Chem. Dose, mg/L} \times 8.34\ \text{lb/gal} \quad (8.15)$$

To calculate the pounds per minute chemical required, we use equation 8.16.

$$\text{Chemical, lb/min} = \frac{\text{Chemical, lb/day}}{1440\ \text{min/day}} \quad (8.16)$$

Example 8.19

Problem:

Jar tests indicate that the optimum lime dosage is 200 mg/L. If the flow to be treated is 4 MGD, what should be the chemical feeder setting in pounds per day and pounds per minute?

Solution:

Calculate the pounds per day feed rate using equation 8.15:

$$\text{Feed Rate, lb/day} = (\text{Flow Rate, MGD})\ (\text{Chemical Dose, mg/L})\ (8.34\ \text{lb/gal})$$
$$= (200\ \text{mg/L})\ (4.0\ \text{MGD})\ (8.34\ \text{lb/gal})$$
$$= 6672\ \text{lb/day}$$

Convert this feed rate to pounds per minute:

$$= \frac{6672 \text{ lb/day}}{1440 \text{ min/day}}$$

$$= 4.6 \text{ lb/min}$$

Example 8.20

Problem:

What should the lime dosage setting be, in pounds per day and pounds per hour, if the optimum lime dosage has been determined to be 125 mg/L and the flow to be treated is 1.1 MGD?

The pounds per day feed rate for lime is:

$$\text{Lime, lb/day} = (\text{Lime, mg/L}) \; (\text{Flow, MGD}) \; (8.34 \text{ lb/gal})$$

$$= (125 \text{ mg/L}) \; (1.1 \text{ MGD}) \; (8.34 \text{ lb/day})$$

$$= 1147 \text{ lb/day}$$

Convert this to pounds per minute feed rate, as follows:

$$= \frac{1147 \text{ lb/day}}{24 \text{ hr/day}}$$

$$= 48 \text{ lb/hr}$$

8.8   ION EXCHANGE CAPACITY

An ion exchange softener is a common alternative to the use of lime and soda ash for softening water. Natural water sources contain dissolved minerals that dissociate in water to form charged particles called ions. Of main concern are the positively charged ions of calcium, magnesium, and sodium, and bicarbonate, sulfate, and chloride are the normal negatively charged ions of concern. An ion exchange medium, called resin, is a material that will exchange a hardness-causing ion for another one that does not cause hardness, hold the new ion temporarily, and then release it when a regenerating solution is poured over the resin. The removal capacity of an exchange resin is generally reported as grains of hardness removal per cubic foot of resin. To calculate the removal capacity of the softener, we use equation 8.17.

$$\text{Exchange Cap., grains} = (\text{Removal Cap., grains/cu ft}) \; (\text{Media Vol., cu ft}) \quad (8.17)$$

Example 8.21

Problem:

The hardness removal capacity of an exchange resin is 24,000 grains/cu ft. If the softener contains a total of 70 cu ft of resin, what is the total exchange capacity (grains) of the softener?

Solution:

$$\text{Exchange Cap., grains} = (\text{Removal Cap., grains/cu ft}) \ (\text{Media Vol., cu ft})$$
$$= (22,000 \text{ grains/cu ft}) \ (70 \text{ cu ft})$$
$$= 1,540,000 \text{ grains}$$

Example 8.22

Problem:

An ion exchange water softener has a diameter of 7 ft. The depth of resin is 5 ft. If the resin has a removal capacity of 22 kg/cu ft, what is the total exchange capacity of the softener (in grains)?

Solution:

Before the exchange capacity of a softener can be calculated, the cubic foot resin volume must be known:

$$\text{Vol., cu ft} = (0.785) \ (D^2) \ (\text{Depth, ft})$$
$$= (0.785) \ (7 \text{ ft}) \ (7 \text{ ft}) \ (5 \text{ ft})$$
$$= 192 \text{ cu ft}$$

Calculate the exchange capacity of the softener:

$$\text{Exchange Cap., grains} = (\text{Removal Cap., grains/ft}) \ (\text{Media Vol., cu ft})$$
$$= (22,000 \text{ grains/cu ft}) \ (192 \text{ cu ft})$$
$$= 4,224,000 \text{ grains}$$

8.9 WATER TREATMENT CAPACITY

To calculate when the resin must be regenerated (based on volume of water treated), we know the exchange capacity of the softener and the hardness of the water. Equation 8.18 is used for this calculation:

$$\text{Water Treatment Cap., gal} = \frac{\text{Exchange Capacity, grains}}{\text{Hardness, grains/gallon}} \qquad (8.18)$$

Example 8.23

Problem:

An ion exchange softener has an exchange capacity of 2,445,000 grains. If the hardness of the water to be treated is 18.6 grains/gal, how many gallons of water can be treated before regeneration of the resin is required?

Solution:

$$\text{Water Treatment Capacity, gal} = \frac{\text{Exchange Capacity, grains}}{\text{Hardness, grains/gallon}}$$

$$= \frac{2,455,000 \text{ grains}}{18.6 \text{ gpg}}$$

$$= 131,989 \text{ gallons water treated}$$

Example 8.24

Problem:

An ion exchange softener has an exchange capacity of 5,500,000 grains. If the hardness of the water to be treated is 14.8 grains/gal, how many gallons of water can be treated before regeneration of the resin is required?

Solution:

$$\text{Water Treatment Capacity, gal} = \frac{\text{Exchange Capacity, grains}}{\text{Hardness, grains/gallon}}$$

$$= \frac{5,500,000 \text{ grains}}{14.8 \text{ gpg}}$$

$$= 371,622 \text{ gallons water treated}$$

Example 8.25

Problem:

The hardness removal capacity of an ion exchange resin is 25 kg/cu ft. The softener contains a total of 160 cu ft of resin. If the water to be treated contains 14 gpg hardness, how many gallons of water can be treated before regeneration of the resin is required?

Solution:

Both the water hardness and the exchange capacity of the softener must be determined before the gallons water can be calculated.

$$\text{Exchange Capacity, grains} = (\text{Removal Capacity, grains/cu ft}) \ (\text{Media Volume, cu ft})$$

$$= (25,000 \text{ grains/cu ft}) \ (160 \text{ cu ft})$$

$$= 4,000,000 \text{ grains}$$

Calculate the gallons water treated:

$$\text{Water Treatment Capacity, gal} = \frac{4,000,000 \text{ grains}}{14.0 \text{ gpg}}$$

$$= 285,714 \text{ gallons water treated}$$

8.10   TREATMENT TIME CALCULATION (UNTIL REGENERATION REQUIRED)

After calculating the total number of gallons water to be treated (before regeneration), we can also calculate the operating time required to treat that amount of water. Equation 8.19 is used to make this calculation.

$$\text{Operating Time, hr} = \frac{\text{Water Treated, gal}}{\text{Flow Rate, gph}} \qquad (8.19)$$

Example 8.26

Problem:

An ion exchange softener can treat a total of 642,000 gal before regeneration is required. If the flow rate treated is 25,000 gph, how many hours of operation are there before regeneration is required?

Solution:

$$\text{Operating Time, hrs} = \frac{\text{Water Treated, gal}}{\text{Flow Rate, gph}}$$

$$= \frac{642,000 \text{ gal}}{25,000 \text{ gph}}$$

$$= 25.7 \text{ hrs of operation before regeneration}$$

Example 8.27

Problem:

An ion exchange softener can treat a total of 820,000 gal of water before regeneration of the resin is required. If the water is to be treated at a rate of 32,000 gph, how many hours of operation are there until regeneration is required?

Solution:

$$\text{Operating Time, hrs} = \frac{\text{Water Treatment, gal}}{\text{Flow Rate, gph}}$$

$$= \frac{820,000 \text{ gal}}{32,000 \text{ gph}}$$

$$= 25.6 \text{ hr of operation before regeneration}$$

8.11   SALT AND BRINE REQUIRED FOR REGENERATION

When calcium and magnesium ions replace the sodium ions in the ion exchange resin, the resin can no longer remove the hardness ions from the water. When this occurs, pumping a concentrated solution (10–14% sodium chloride solution) on the resin must regenerate the resin. When the resin is completely recharged with sodium ions, it is ready for softening again. Typically, the salt dosage required to prepare the brine solution ranges from 5 tp 15 lb salt/cu ft resin. Equation 8.20 is used to calculate salt required (pounds, lb), and equation 8.21 is used to calculate brine (gallons).

$$\text{Salt Required, lb} = \frac{\left(\text{Salt Req., lb/kgrains removed}\right)}{\left(\text{Hard. removed, kgrains}\right)} \qquad (8.20)$$

$$\text{Brine (gal)} = \frac{\text{Salt Required, lb}}{\text{Brine Solution, lb salt/gal brine}} \qquad (8.21)$$

To determine the brine solution pounds salt per gallon brine factor used in equation 8.21, we must refer to the salt solutions table shown in the following.

8.12   SALT SOLUTIONS TABLE

% NaCl	NaCl (Lb/Gal)	NaCl/Cu Ft (Lb)
10	0.874	6.69
11	0.990	7.41
12	1.09	8.14
13	1.19	8.83
14	1.29	9.63
15	1.39	10.4

Example 8.28

Problem:

An ion exchange softener removes 1,310,000 grains hardness from the water before the resin must be regenerated. If 0.3 lb salt are required for each kilograin removed, how many pounds of salt will be required for preparing the brine to be used in resin regeneration?

Solution:

$$\text{Salt Required, lb} = \left(\text{Salt Required, lb/1000 grains}\right)\left(\text{Hardness Removed, kg}\right)$$
$$= \left(0.3 \text{ lb salt/kilograins removed}\right)\left(1310 \text{ kilograins}\right)$$
$$= 393 \text{ lb salt required}$$

Example 8.29

Problem:

A total of 430 lb salt will be required to regenerate an ion exchange softener. If the brine solution is to be a 12% (see salt solutions table to determine the pounds salt per gallon brine for a 12% brine solution) brine solution, how many gallons brine will be required?

Solution:

$$\text{Brine, gal} = \frac{\text{Salt Required, lb}}{\text{Brine Solution, lb salt/gal brine}}$$

$$= \frac{430 \text{ lb salt}}{1.09 \text{ lb salt/gal brine}}$$

$$= 394 \text{ gal of } 12\% \text{ brine}$$

Thus, it takes 430 lb salt to make up a total of 394 gal brine, which will result in the desired 12% brine solution.

9 Water Treatment Practice Calculations: Basic Math Problems

9.1 DECIMAL OPERATIONS

1. $90.5 \times 7.3 =$

2. $9.556 \times 1.03 =$

3. $13 \div 14.3 =$

4. $8.2 \div 0.96 =$

5. $2 \div 0.053 =$

6. Convert 3/4 into a decimal.

7. Convert 1/6 into a decimal.

DOI: 10.1201/9781003354307-9

8. Convert 3/8 into a decimal.

9. Convert 0.13 into a fraction.

10. Convert 0.9 into a fraction.

11. Convert 0.75 into a fraction.

12. Convert 0.245 into a fraction.

9.2   PERCENTAGE CALCULATIONS

13. Convert 15/100 into a percent.

14. Convert 122/100 into a percent.

15. Convert 1.66 into a percent.

16. Convert 4/7 into a percent.

17. A 100% decline from 66 leaves us with how much?

9.3 FIND X

18. If $x - 6 = 2$, how much is x?

19. If $x - 4 = 9$, how much is x?

20. If $x - 8 = 17$, how much is x?

21. If $x + 10 = 15$, how much is x?

22. Find x when $x/3 = 2$.

24. Solve for x when $x/4 = 10$.

25. If $4x = 8$, how much is x?

26. If $6x = 15$, how much is x?

26. If $x + 10 = 2$, how much is x?

27. Find x if $x - 2 = -5$

28. If $x + 4 = -8$, how much is x?

29. If $x - 10 = -14$, how much is x?

30. $0.5x - 1 = -6$. Find x.

31. If $9x + 1 = 0$, how much is x?

32. How much is x^2 if $x = 6$?

33. If $x = 3$, how much is x^4?

34. If $x = 10$, how much is x^0?

9.4 RATIO AND PROPORTION

35. If an employee was out sick on 6 of 96 workdays, what is her ratio of sick days to days worked?

36. Find x when $2:x = 5:15$.

9.5   AREA OF RECTANGLES

37. How much is the area of the rectangle shown in Figure 37?

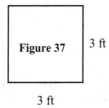

3 ft

3 ft

38. How much is the area of the rectangle shown in Figure 38?

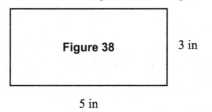

3 in

5 in

39. How much is the area of the rectangle shown in Figure 39?

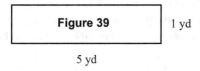

1 yd

5 yd

9.6   CIRCUMFERENCE AND AREA OF CIRCLES

40. Find the circumference of a circle whose diameter is 14 ft.

41. If the circumference of a circle is 8 in, how much is the diameter?

42. If the radius of a circle is 7 in, what is its area?

43. If the diameter of a circle is 10 in, what is its area?

9.7   WATER TREATMENT PROBLEMS

1. The static water level for a well is 91 ft. If the pumping water level is 98 ft, what is the well drawdown?

2. The static water level for a well is 110 ft. The pumping water level is 125 ft. What is the well drawdown?

3. Before the pump is started, the water level is measured at 144 ft. The pump is then started. If the pumping water level is determined to be 161 ft, what is the well drawdown?

4. The static water level of a well is 86 ft. The pumping water level is determined using the sounding line. The air pressure applied to the sounding line is 3.7 psi, and the length of the sound line is 112 ft. What is the drawdown?

5. A sounding line is used to determine the static water level for a well. The air pressure applied to 4.6 psi and the length of the sounding line is 150 ft. If the pumping water level is 171 ft, what is the drawdown?

6. If the well yield is 300 gpm and the drawdown is measured to be 20 ft, what is the specific capacity?

7. During a 5 min well yield test, a total of 420 gal was pumped from the well. What is the well yield in gallons per minute?

8. Once the drawdown of a well stabilized, it was determined that the well produced 810 gal during a 5 min pumping test. What is the well yield in gallons per minute?

9. During a test for well yield, a total of 856 gal was pumped from the well. If the well yield test lasted 5 min, what was the well yield in gallons per minute? In gallons per hour?

10. A bailer is used to determine the approximate yield of a well. The bailer is 12 ft long and has a diameter of 12 in. If the bailer is placed in the well and removed a total of 12 times during a 5 min test, what is the well yield in gallons per minute?

11. During a 5 min well yield test, a total of 750 gal of water was pumped from the well. At this yield, if the pump is operated a total of 10 hr each day, how many gallons of water are pumped daily?

12. The discharge capacity of a well is 200 gpm. If the drawdown is 28 ft, what is the specific yield in gallons per minute per foot of drawdown?

13. A well produces 620 gpm. If the drawdown for the well is 21 ft, what is the specific yield in gallons per minute per foot of drawdown?

14. A well yields 1,100 gpm. If the drawdown is 41.3 ft, what is the specific yield in gallons per minute per foot of drawdown?

15. The specific yield of a well is listed as 33.4 gpm/ft. If the drawdown for the well is 42.8 ft, what is the well yield in gallons per minute?

16. A new well is to be disinfected with chlorine at a dosage of 40 mg/L. If the well casing diameter is 6 in and the length of the water-filled casing is 140 ft, how may pounds of chlorine will be required?

17. A new well with a casing diameter of 12 in is to be disinfected. The desired chlorine dosage is 40 mg/L. If the casing is 190 ft long and the water level in the well is 81 ft from the top of the well, how many pounds of chlorine will be required?

18. An existing well has a total casing length of 210 ft. The top 180 ft of casing has a 12 in diameter, and the bottom 40 ft of the casing has an 8 in diameter. The water level is 71 ft from the top of the well. How many pounds of chlorine will be required if a chlorine dosage of 110 mg/L is desired?

19. The water-filled casing of a well has a volume of 540 gal. If 0.48 lb of chlorine were used in disinfection, what was the chlorine dosage in milligrams per liter?

20. A total of 0.09 lb of chlorine is required for the disinfection of a well. If sodium hypochlorite (5.25% available chlorine) is to be used, how many fluid ounces of sodium hypochlorite are required?

21. A new well is to be disinfected with calcium hypochlorite (65% available chlorine). The well casing diameter is 6 in, and the length of the water-filled

casing is 120 ft. If the desired chlorine dosage is 50 g/L, how many ounces (dry measure) of calcium hypochlorite will be required?

22. How many pounds of chloride of lime (25% available chlorine) will be required to disinfect a well if the casing is 18 in in diameter and 200 ft long, with a water level at 95 ft from the top of the well? The desired chlorine dosage is 100 mg/L.

23. The water-filled casing of a well has a volume of 240 gal. How many fluid ounces of sodium hypochlorite (5.25% available chlorine) are required to disinfect the well if a chlorine concentration of 60 mg/L is desired?

24. The pressure gauge reading at a pump discharge head is 4.0 psi. What is this discharge head expressed in feet?

25. The static water level of a well is 94 ft. The well drawdown is 24 ft. If the gauge reading at the pump discharge head is 3.6 psi, what is the field head?

26. The pumping water level for a well is 180 ft. The discharge pressure measured at the pump discharge head is 4.2 psi. If the pump capacity is 800 gpm, what is the water horsepower?

27. The pumping water level for a well is 200 ft. The pump discharge head is 4.4 psi. If the pump capacity is 1,000 gpm, what is the water horsepower?

28. The bowl head of a vertical turbine pump is 184 ft, and the bowl efficiency is 83%. If the capacity of the vertical turbine pump is 700 gpm, what is the bowl horsepower?

29. A vertical turbine pump has a bowl horsepower of 59.5 bhp. The shaft is 1 1/4 in in diameter and rotates at a speed of 1,450 rpm. If the shaft is 181 ft long, what is the field brake horsepower? Shaft friction loss = 0.67.

30. The field brake horsepower for a deep well turbine pump is 58.3 bhp. The thrust bearing loss is 0.5 hp. If the motor efficiency provided by the manufacturer is 90%, what is the horsepower input to the motor?

31. The total brake horsepower for a deep well turbine pump is 56.4 bhp. If the water horsepower is 45 whp, what is the field efficiency?

32. The total brake horsepower for a pump is 55.7 bhp. If the motor is 90% efficient and the water horsepower is 43.5 whp, what is the overall efficiency of the unit?

33. A pond has an average length of 400 ft, an average width of 110 ft, and an estimated average depth of 14 ft. What is the estimated volume of the pond in gallons?

34. A pond has an average length of 400 ft and an average width of 110 ft. If the maximum depth of the pond is 30 ft, what is the estimated gallon volume of the pond?

35. A pond has an average length of 200 ft, an average width of 80 ft, and an average depth of 12 ft. What is the acre-feet volume of the pond?

36. A small pond has an average length of 320 ft, an average width of 170 ft, and a maximum depth of 16 ft. What is the acre-feet volume of the pond?

37. For algae control in a reservoir, a dosage of 0.5 mg/L copper is desired. The reservoir has a volume of 20 MG. How many pounds of copper sulfate pentahydrate (25% available copper) will be required?

38. The desired copper dosage in a reservoir is 0.5 mg/L. The reservoir has a volume of 62 ac-ft. How many pounds of copper sulfate pentahydrate (25% available copper) will be required?

39. A pond has a volume of 38 ac-ft. If the desired copper sulfate dosage is 1.1 lb $CuSO_4$/ac-ft, how many pounds of copper sulfate will be required?

40. A pond has an average length of 250 ft, an average width of 75 ft, and an average depth of 10 ft. If the desired dosage is 0.8 lb copper sulfate/ac-ft, how many pounds of copper sulfate will be required?

41. A storage reservoir has an average length of 500 ft and an average width of 100 ft. If the desired copper sulfate dosage is 5.1 lb $CuSO_4$/ac, how many pounds of copper sulfate will be required?

42. The static water level for a well is 93.9 ft. If the pumping water level is 131.5 ft, what is the drawdown?

43. During a 5 min well yield test, a total of 707 gal was pumped from the well. What is the well yield in gallons per minute? In gallons per hour?

44. A bailer is used to determine the approximate yield of a well. The bailer is 12 ft long and has a diameter of 12 in. If the bailer is placed in the well and removed a total of eight times during a 5 min test, what is the well yield in gallons per minute?

45. The static water level in a well is 141 ft. The pumping water level is determined using the sounding line. The air pressure applied to the sounding line is 3.5 psi, and the length of the sounding line is 167 ft. What is the drawdown?

46. A well produces 610 gpm. If the drawdown for the well is 28 ft, what is the specific yield in gallons per minute per foot of drawdown?

47. A new well is to be disinfected with a chlorine dose of 55 mg/L. If the well casing diameter is 6 in and the length of the water-filled casing is 150 ft, how many pounds of chlorine will be required?

48. During a 5 min well yield test, a total of 780 gal of water was pumped from the well. At this yield, if the pump is operated a total of 8 hr each day, how many gallons of water are pumped daily?

49. The water-filled casing of a well has a volume of 610 gal. If 0.47 lb of chlorine were used for disinfection, what was the chlorine dosage in milligrams per liter?

50. An existing well has a total casing length of 230 ft. The tope 170 ft of casing has a 12 in diameter, and the bottom 45 ft of casing has an 8 in diameter. The water level is 81 ft from the top of the well. How many pounds of chlorine will be required if a chlorine dosage of 100 mg/L is desired?

51. A total of 0.3 lb of chlorine is required for the disinfection of a well. If sodium hypochlorite is to be used (5.25% available chlorine), how many fluid ounces of sodium hypochlorite are required?

52. The pressure gauge reading at a pump discharge head is 4.5 psi. What is this discharge head expressed in feet?

53. The static water level of a well is 95 ft. The well drawdown is 25 ft. If the gauge reading at the pump discharge head is 3.6 psi, what is the field head?

54. The pumping water level for a well is 191 ft. The discharge pressure measured at the pump discharge head is 4.1 psi. If the pump capacity is 850 gpm, what is the horsepower?

55. A deep well vertical turbine pump delivers 800 gpm. The bowl head is 175 ft, and the bowl efficiency is 80%. What is the bowl horsepower?

56. The field brake horsepower for a deep well turbine pump is 47.8 bhp. The thrust bearing loss is 0.8 hp. If the motor efficiency provided by the manufacturer is 90%, what is the horsepower input to the motor?

57. The total brake horsepower for a deep well turbine is 57.4 bhp. If the water horsepower is 45.6 whp, what is the field efficiency?

58. The total brake horsepower for a pump is 54.7 bhp. If the motor is 90% efficient and the water horsepower is 44.6 whp, what is the overall efficiency of the unit?

59. The desired copper dosage at a reservoir is 0.5 mg/L. The reservoir has a volume 53 ac-ft. How many pounds of copper sulfate pentahydrate (25% available copper) will be required?

60. A storage reservoir has an average length of 440 ft and an average width of 140 ft. If the desired copper sulfate dosage is 5.5 lb copper sulfate/ac, how many pounds of copper sulfate will be required?

61. A flash mix chamber is 4 ft wide, 5 ft long, with water to a depth of 3 ft. What is the gallon volume of water in the flash mix chamber?

62. A flocculation basin is 50 ft long, 20 ft wide, with water to a depth of 8 ft. What is the volume of water in the basin (in gallons)?

63. A flocculation basin is 40 ft long, 16 ft wide, with water to a depth of 8 ft. How many gallons of water are in the basin?

64. A flash mix chamber is 5 ft square, with water to a depth of 42 in. What is the volume of water in the flash mixing chamber (in gallons)?

65. A flocculation basin is 25 ft wide, 40 ft long, and contains water to a depth of 9 ft, 2 in. What is the volume of water (in gallons) in the flocculation basin?

66. The flow to a flocculation basin is 3,625,000 gpd. If the basin is 60 ft long, 25 ft wide, with water to a depth of 9 ft, what is the detention time (in minutes) of the flocculation basin?

67. A flocculation basin is 50 ft long, 20 ft wide, and has a water level of 8 ft. What is the detention time (in minutes) in the basin if the flow to the basin is 2.8 MGD?

68. A flash mix chamber 6 ft long, 5 ft wide, and 5 ft deep receives a flow of 9 MGD. What is the detention time in the chamber (in seconds)?

69. A flocculation basin is 50 ft long, 20 ft wide, and has a water depth of 10 ft. If the flow to the basin is 2,250,000 gpd, what is the detention time in minutes?

70. A flash mix chamber is 4 ft square, with a water depth of 42 in. If the flash mix chamber receives a flow of 3.25 MGD, what is the detention time in seconds?

71. The desired dry alum dosage, as determined by the jar test, is 10 mg/L. Determine the pounds per day setting on a dry alum feeder if the flow is 3,450,000 gpd.

72. Jar tests indicate that the best polymer dose for a water sample is 12 mg/L. If the flow to be treated is 1,660,000 gpd, what should the dry chemical feed setting be in pounds per day?

73. Determine the desired pounds per day setting on a dry alum feeder if jar tests indicate an optimum dose of 10 mg/L and the flow to be treated is 2.66 MGD.

74. The desired dry alum dose is 9 mg/L, as determined by a jar test. If the flow to be treated is 940,000 gpd, how many pounds per day dry alum will be required?

75. A flow of 4.10 MGD is to be treated with a dry polymer. If the desired dose is 12 mg/L, what should the dry chemical feeder setting be (in pounds per day)?

76. Jar tests indicate that the best alum dose for a unit process is 7 mg/L. The flow to be treated is 1.66 MGD. Determine the gallons per day setting for

the alum solution feeder if the liquid alum contains 5.24 lb of alum per gallon of solution.

77. The flow to a plant is 3.43 MGD. Jar testing indicates that the optimum alum dose is 12 mg/L. What should the gallons per day setting be for the solution feeder if the alum solution is 55% solution?

78. Jar tests indicate that the best alum dose for a unit process is 10 mg/L. The flow to be treated is 4.13 MGD. Determine the gallons per day setting for the alum solution feeder if the liquid alum contains 5.40 lb of alum per gallon of solution.

79. Jar tests indicate that the best liquid alum dose for a unit process is 11 mg/L. The flow to be treated is 880,000 gpd. Determine the gallons per day setting for the liquid alum chemical feeder if the liquid alum is a 55% solution.

80. A flow of 1,850,000 gpd is to be treated with alum. Jar tests indicate that the optimum alum dose is 10 mg/L. If the liquid alum contains 640 mg alum/mL solution, what should be the gallons per day setting for the alum solution feeder?

81. The desired solution feed rate was calculated to be 40 gpd. What is this feed rate expressed as milliliters per minute?

82. The desired solution feed rate was calculated to be 34.2 gpd. What is this feed rate expressed as milliliters per minute?

83. The optimum polymer dose has been determined to be 10 mg/L. The flow to be treated is 2,880,000 gpd. If the solution to be used contains 55% active polymer, what should the solution chemical feeder setting be in milliliters per minute?

84. The optimum polymer dose for a 2,820,000 gpd flow has been determined to be 6 mg/L. If the polymer solution contains 55% active polymer, what should the solution chemical feeder setting be in milliliters per minute? Assume the polymer solution weighs 8.34 lb/day.

85. Jar tests indicate that the best alum dose for a unit process is 10 mg/L. The liquid alum contains 5.40 alum per gallon of solution. What should the setting be on the solution chemical feeder (in milliliters per minute) when the flow to be treated is 3.45 MGD?

86. If 140 g of dry polymer are dissolved in 16 gal of water, what is the percent strength of the solution (1 g = 0.0022 lb)?

87. If a total of 22 oz of dry polymer is added to 24 gal of water, what is the percent strength (by weight) of the polymer solution?

88. How many gallons of water must be added to 2.1 lb dry alum to make a 0.8% solution?

89. An 11% liquid polymer is to be used in making up a polymer solution. How many pounds of liquid polymer should be mixed with water to produce 160 lb of a 0.5% polymer solution?

90. An 8% polymer solution is used in making up a solution. How many gallons of liquid polymer should be added to the water to make up 50 gal of a 0.2% polymer solution? The liquid polymer has a specific gravity of 1.3. Assume the polymer solution has a specific gravity of 1.0.

91. How many gallons of an 11% liquid polymer should be mixed with water to produce 80 gal of a 0.8% polymer solution? The density of the polymer liquid is 10.1 lb/gal. Assume the density of the polymer solution is 8.34 lb/gal.

92. If 32 lb of a 10% strength solution are mixed with 66 lb of a 0.5% strength solution, what is the percent strength of the solution mixture?

93. If 5 gal of a 15% strength solution are added to 40 gal of a 0.20% strength solution, what is the percent strength of the solution mixture? Assume the 15% strength solution weighs 11.2 lb/gal and the 0.20% solution weighs 8.34 lb/gal.

94. If 12 gal of a 12% strength solution is mixed with 50 gal of a 0.75% strength solution, what is the percent strength of the solution mixture? Assume the 12% solution weighs 10.5 lb/gal and the 0.75% solution weighs 8.40 lb/gal.

95. Calculate the actual chemical feed rate, in pounds per day, if a container is placed under a chemical feeder and a total of 2.3 lb is collected during a 30 min period.

96. Calculate the actual chemical feed rate, in pounds per day, if a container is placed under a chemical feeder and a total of 42 oz is collected during a 45 min period.

97. To calibrate a chemical feeder, a container is first weighed (14 oz) then placed under the chemical feeder. After 30 min, the container is weighed again. If the weight of the bucket with chemical is 2.4 lb, what is the actual chemical feed rate in pounds per day?

98. A chemical feeder is to be calibrated. The container to be used to collect chemical is placed under the chemical feeder and weighed (0.6 lb). After 30 min, the weight of the bucket is found to be 2.8 lb. Based on this test, what is the actual chemical feed rate in pounds per day?

99. During a 24 hr period, a flow of 1,920,000 gpd water is treated. If a total of 42 lb of polymer is used for coagulation during that 24 hr period, what is the polymer dosage in milligrams per liter?

100. A calibration test is conducted for a solution chemical feeder. During a 24 hr period, the solution feeder delivers a total of 70 gal solution. The polymer solution is a 1.6 % solution. What is the pounds per day solution feed rate? Assume the polymer solution weighs 8.34 lb/gal.

101. A calibration test is conducted for a solution chemical feeder. During a 5 min test, the pump delivered 590 mL of a 1.2% polymer solution. The specific gravity of the polymer solution is 1.09. What is the polymer dosage rate in pounds per day?

102. During a 5 min calibration test for a solution chemical feeder, the solution feeder delivers a total of 725 mL. The polymer solution is a 1.2% solution. What is the pounds per day polymer feed rate? Assume the polymer solution weighs 8.34 lb/gal.

103. A solution chemical feeder delivered 950 mL solution during a 5 min calibration test. The polymer solution is a 1.4% strength solution. What is the polymer dosage rate in pounds per day? Assume the polymer solution weighs 8.34 lb/gal.

104. If 1,730 mL of a 1.9% polymer solution are delivered during a 10 min calibration test and the polymer solution has a specific gravity of 1.09, what is the polymer feed rate in pounds per day?

105. A pumping rate calibration test is conducted for a 5 min period. The liquid level in the 4 ft diameter solution tank is measured before and after the test. If the level drops 4 in during the 5 min test, what is the gallons per minute pumping rate?

106. During a 15 min pumping rate calibration test, the liquid level in the 4 ft diameter solution tank drops 4 in. What is the pumping rate in gallons per minute?

107. The liquid level in a 3 ft diameter solution tank drops 3 in during a 10 min pumping rate calibration test. What is the gallons per minute pumping rate? Assuming a continuous pumping rate, what is the gallons per day pumping rate?

108. During a 15 min pumping rate calibration test, the solution level in the 3 ft diameter chemical tank dropped 2 in. Assume the pump operates at the rate during the next 24 hr. If the polymer solution is 1.3% solution, what is the pounds per day polymer feed? Assume the polymer solution weighs 8.34 lb/gal.

109. The level in a 4 ft diameter chemical tank drops 2 in during a 30 min pumping rate calibration test. The polymer solution is 1.45% solution. If the pump operates at the same rate for 24 hr, what is the polymer feed in pounds per day?

110. The amount of chemical used for each day during a week is given in the following. Based on this data, what was the average pounds per day chemical use during the week?

Monday: 81 lb/day   Friday: 79 lb/day
Tuesday: 73 lb/day   Saturday: 80 lb/day
Wednesday: 74 lb/day   Sunday: 82 lb/day
Thursday: 66 lb/day

111. The average chemical use at a plant is 90 lb/day. If the chemical inventory in stock is 2,200 lb, how many days' supply is this?

112. The chemical inventory in stock is 889 lb. If the average chemical use at a plant is 58 lb/day, how many days' supply is this?

113. The average gallons of polymer solution used each day at a treatment plant is 88 gpd. A chemical feed tank has a diameter of 3 ft and contains solution to a depth of 3 ft, 4 in. How many days' supply are represented by the solution in the tank?

114. Jar tests indicate that the optimum polymer dose for a unit process is 2.8 mg/L. If the flow to be treated is 1.8 MGD, how many pounds of dry polymer will be required for a 30-day period?

115. A flash mix chamber 4 ft long, 4 ft wide, with a 3 ft water depth receives a flow of 6.1 MGD. What is the detention time in the chamber (in seconds)?

116. A flocculation basin is 50 ft long, 20 ft wide, and has a water depth of 9 ft. What is the volume of water in the basin (in gallons)?

117. The desired dry alum dosage, as determined by jar testing, is 9 mg/L. Determine the pounds per day setting on a dry alum feeder if the flow is 4.35 MGD.

118. The flow to a plant is 3.15 MGD. Jar testing indicates that the best alum dose is 10 mg/L. What should the gallons per day setting be for the solution feeder if the alum solution is a 50% solution? Assume the alum solution weighs 8.34 lb/gal.

119. A flash mix chamber is 4 ft square, with a water depth of 2 ft. What is the gallon volume of water in this chamber?

120. The desired solution feed rate was calculated to be 45 gpd. What is this feed rate expressed as milliliters per minute?

121. A flocculation basin is 40 ft long, 20 ft wide, and has a water depth of 9 ft, 2 in. If the flow to the basin is 2,220,000 gpd, what is the detention time in minutes?

122. The optimum polymer dose has been determined to be 8 mg/L. The flow to be treated is 1,840,000 gpd. If the solution to be used contains 60% active polymer, what should the solution chemical feeder setting be, in milliliters per minute? The polymer solution weighs 10.2 lb/gal.

123. The desired solution feed rate was calculated to be 180 gpd. What is this feed rate expressed as milliliters per minute?

124. Determine the desired pounds per day setting on a dry alum feeder if jar tests indicate an optimum dose of 6 mg/L and the flow to be treated is 925,000 gpd.

125. How many gallons of water must be added to 2.7 lb of dry alum to make a 1.4% solution?

126. If 25 lb of a 16% strength solution are mixed with 140 lb of a 0.6% strength solution, what is the percent strength of the solution mixture?

127. Calculate the chemical feed rate, in pounds per day, if a container is placed under a chemical feeder and a total of 4 lb chemical is collected during a 30 min period.

128. A chemical feeder is to be calibrated. The container to be used to collect chemical is placed under the chemical feeder and weighed (2.0 lb). After 30 min, the weight of the container and chemical is found to be 4.2 lb. Based on this test, what is the chemical feed rate, in pounds per day?

129. If 190 g of dry polymer are dissolved in 25 gal of water, what is the percent strength (by weight) of the solution (1 g = 0.0022 lb)?

130. During a 5 min calibration test for a solution chemical feeder, the feeder delivers a total of 760 mL. The polymer solution is a 2.0% solution. What is the pounds per day polymer feed rate? Assume the polymer solution weighs 8.34 lb/gal.

131. Jar tests indicate that the best alum dose for a unit process is 14 mg/L. The flow to be treated is 4.2 MD. Determine the gallons per day setting for the alum solution feeder if the liquid alum contains 5.66 lb of alum per gallon of solution.

132. A 10% liquid polymer is to be used in making up a polymer solution. How many pounds of liquid polymer should be mixed with water to produce 210 lb of a 0.8% polymer solution?

133. How many pounds of a 60% polymer solution and water should be mixed together to form 175 lb of a 1% solution?

134. During a 10 min pumping rate calibration test, the liquid level in the 4 ft diameter solution tank drops 3 in. What is the pumping rate in gallons per minute?

135. How many gallons of a 10% liquid polymer should be mixed with water to produce 80 gal of a 0.6% polymer solution? The density of the polymer liquid is 10.2 lb/gal. Assume the density of the polymer solution is 8.34 lb/gal.

136. A calibration test is conducted of a solution chemical feeder. During a 5 min test, the pump delivered 710 mL of a 0.9% polymer solution. The specific gravity of the polymer solution is 1.3. What is the polymer dosage rate in pounds per day?

137. Jar tests indicate that the best polymer dose for a unit process is 6 mg/L. If the flow to be treated is 3.7 MGD, at this rate how many pounds of dry polymer will be required for a 30-day period?

138. The chemical inventory in stock is 550 lb. If the average chemical use at the plant is 80 lb/day, how many days' supply is this?

139. A sedimentation basin is 70 ft long and 30 ft wide. If the water depth is 14 ft, what is the volume of water in the tank in gallons?

140. A circular clarifier has a diameter of 80 ft. If the water depth is 12 ft, how many gallons of water are in the tank?

141. A sedimentation tank is 70 ft long, 20 ft wide, and has water to a depth of 10 ft. What is the volume of water in the tank in gallons?

142. A sedimentation basin is 40 ft long and 25 ft wide. When the basin contains a total of 50,000 gal, what would be the water depth?

143. A circular clarifier is 75 ft in diameter. If the water depth is 10 ft, 5 in, what is the volume of water in the clarifier in gallons?

144. A rectangular sedimentation basin is 70 ft long, 25 ft wide, and has water to a depth of 10 ft. The flow to the basin is 2,220,000 gpd. Calculate the detention time in hours for the sedimentation basin.

145. A circular clarifier has a diameter of 80 ft and an average water depth of 12 ft. If the flow to the clarifier is 2,920,000 gpd, what is the detention time in hours?

146. A rectangular sedimentation basin is 60 ft long and 20 ft wide, with water to a depth of 10 ft. If the flow to the basin is 1,520,000 gpd, what is the detention time in hours?

147. A circular clarifier has a diameter of 60 ft and an average water depth of 12 ft. What flow rate (in million gallons per day) corresponds to a detention time of 3 hr?

148. A sedimentation basin is 70 ft long and 25 ft wide. The average water depth is 12 ft. If the flow to the basin is 1,740,000 gpd, what is the detention time in the sedimentation basin in hours?

149. A rectangular sedimentation basin is 60 ft long and 25 ft wide. When the flow is 510 gpm, what is the surface overflow rate in gallons per minute per square foot?

150. A circular clarifier has a diameter of 70 ft. If the flow to the clarifier is 1,610 gpm, what is the surface overflow rate in gallons per minute per square foot?

151. A rectangular sedimentation basin receives a flow of 540,000 gpd. If the basin is 50 ft long and 20 ft wide, what is the surface overflow rate in gallons per minute per square foot?

152. A sedimentation basin is 80 ft long and 25 ft wide. To maintain a surface overflow rate of 0.5 gpm/sq ft, what is the maximum flow to the basin in gallons per day?

153. A circular clarifier 60 ft in diameter receives a flow of 1,820,000 gpd. What is the surface overflow rate in gallons per minute per square foot?

154. A sedimentation basin is 80 ft long, 25 ft wide, and operates at a depth of 12 ft. If the flow to the basin is 1,550,000 gpd, what is the mean flow velocity in feet per minute?

155. A sedimentation basin is 70 ft long, 30 ft wide, and operates at a depth of 12 ft. If the flow to the basin is 1.8 MGD, what is the mean flow velocity in feet per minute?

156. A sedimentation basin is 80 ft long and 30 ft wide. The water level is 14 ft. When the flow to the basin is 2.45 MGD, what is the mean flow velocity in feet per minute?

157. The flow to a sedimentation basin is 2,880,000 gpd. If the length of the basin is 70 ft, the width of the basin is 40 ft, and the depth of water in the basin is 10 ft, what is the mean flow velocity in feet per minute?

158. A sedimentation basin 50 ft long and 25 ft wide receives a flow of 910,000 gpd. The basin operates at a depth of 10 ft. What is the mean flow velocity in the basin in feet per minute?

159. A circular clarifier receives a flow of 2,520,000 gpd. If the diameter of the weir is 70 ft, what is the weir loading rate in gallons per minute per gram?

160. A rectangular sedimentation basin has a total of 170 ft of weir. If the flow to the basin 1,890,000 gpd, what is the weir loading rate in gallons per minute per foot?

161. A rectangular sedimentation basin has a total of 120 ft of weir. If the flow over the weir is 1,334,000 gpd, what is the weir loading rate in gallons per minute per foot?

162. A circular clarifier receives a flow of 3.7 MGD. If the diameter of the weir is 70 ft, what is the weir loading rate in gallons per minute per foot?

163. A rectangular sedimentation basin has a total of 160 ft of weir. If the flow over the weir is 1.9 MGD, what is the weir loading rate in gallons per minute per foot?

164. A 100 mL sample of slurry from a solids contact unit is placed in a graduated cylinder and allowed to settle for 10 min. The settled sludge at the bottom of the graduated cylinder after 10 min is 22 mL. What is the percent settled sludge of the sample?

165. A 100 mL sample of slurry from a solids contact unit is placed in a graduated cylinder. After 10 min, a total of 25 mL of sludge settled to the bottom of the cylinder. What is the percent settled of the sample?

166. A percent settled sludge test is conducted on a 100 mL sample of solids contact unit slurry. After 10 min of settling, a total of 15 mL of sludge is found to settle to the bottom of the cylinder. What is the percent settled sludge of the sample?

167. A 100 mL sample of slurry from a solids contact unit is placed in a graduated cylinder. After 10 min, a total of 16 mL of sludge is found to have settled to the bottom of the cylinder. What is the percent settled sludge of the sample?

168. Raw water requires an alum dose of 52 mg/L, as determined by jar testing. If a "residual" 40 mg/L alkalinity must be present in the water to promote

precipitation of the alum added, what is the total alkalinity required in milligrams per liter (1 mg/L alum reacts with 0.45 mg/L alkalinity)?

169. Jar tests indicate that 60 mg/L alum are optimum for a particular raw water unit. If a "residual" 30 mg/L alkalinity must be present to promote complete precipitation of the alum added, what is the total alkalinity required in milligrams per liter (1 mg/L alum reacts with 0.45 mg/L alkalinity)?

170. A total of 40 mg/L alkalinity is required to react with alum and ensure proper precipitation. If the raw water has an alkalinity of 26 mg/L as bicarbonate, how may milligrams per liter alkalinity should be added to the water?

171. A total of 40 mg/L alkalinity is required to react with alum and ensure complete precipitation of the alum added. If the raw water has an alkalinity of 28 mg/L as bicarbonate, how many milligrams per liter alkalinity should be added to the water?

172. A total of 15 mg/L alkalinity must be added to raw water. How many milligrams per liter lime will be required to provide this amount of alkalinity (1 mg/L alum reacts with 0.45 mg/L alkalinity and 1 mg/L alum reacts with 0.45 mg/L lime)?

173. It has been calculated that 20 mg/L alkalinity must be added to a raw water unit. How many milligrams per liter lime will be required to provide this amount of alkalinity (1 mg/L alum reacts with 0.45 mg/L alkalinity and 1 mg/L alum reacts with 0.35 mg/L lime)?

174. Given the following data, calculate the required lime dose in milligrams per liter.

Alum dose required/jar tests: 55 mg/L   1 mg/L alum reacts
                                                 w/0.45 mg/L alk.

Raw water Alkalinity: 35 mg/L   1 mg/L alk. reacts
                                                 w/0.35 mg/L lime

"Residual" Alk. required for precipitation: 30 mg/L

175. The lime dose for a raw water unit has been calculated to be 13.8 mg/L. If the flow to be treated is 2.7 MGD, how many pounds dry lime will be required?

176. The lime dose for a solids contact unit has been calculated to be 12.3 mg/L. If the flow to be treated is 2,400,000 gpd, how many pounds per day lime will be required?

177. The flow to a solids contact clarifier is 990,000 gpd. If the lime dose required is determined to be 16.1 mg/L, how many pounds per day lime will be required?

178. A solids contact clarification unit receives a flow of 2.2 MGD. Alum is to be used for coagulation purposes. If a lime dose of 15 mg/L will be required, how many pounds per day lime is this?

179. A total of 205 lb/day lime will be required to raise the alkalinity of the water passing through a solids-contact clarifier. How many grams per minute lime does this represent (1 lb = 453.6 g)?

180. The lime dose of 110 lb/day is required for a raw water unit feeding a solids-contact clarifier. How much grams per minute lime does this represent (1 lb = 453.6 g)?

181. A lime dose of 12 mg/L is required to raise the pH of a particular water. If the flow to be treated is 900,000 gpd, what grams per minute lime dose will be required (1 lb = 453.6 g)?

182. A lime dose of 14 mg/L is required to raise the alkalinity of a unit process. If the flow to be treated is 2,660,000 gpd, what is the grams per minute lime dose required (1 lb = 452.6 g)?

183. A rectangular sedimentation basin is 65 ft long and 30 ft wide, with water to a depth of 12 ft. If the flow to the basin is 1,550,000 gpd, what is the detention time in hours?

184. A sedimentation basin is 70 ft long and 30 ft wide. If the water depth is 14 ft, what is the volume of water in the tank in gallons?

185. A sedimentation basin 65 ft long and 25 ft wide operates at a depth of 12 ft. If the flow to the basin is 1,620,000 gpd, what is the mean flow velocity in feet per minute?

186. A rectangular sedimentation basin receives a flow of 635,000 gpd. If the basin is 40 ft long and 25 ft wide, what is the surface overflow rate in gallons per minute per square foot?

187. A circular clarifier has a diameter of 70 ft. If the water depth is 14 ft, how many gallons of water are in the tank?

188. A rectangular sedimentation basin has a total of 180 ft of weir. If the flow to the basin is z2,220,000 gpd, what is the weir loading rate in gallons per minute per foot?

189. A circular clarifier has a diameter of 60 ft and an average water depth of 10 ft. If the flow to the clarifier is 2.56 MGD, what is the detention time in hours?

190. A sedimentation basin 55 ft long and 30 ft wide operates at a depth of 12 ft. If the flow to the basin is 1.75 MGD, what is the mean flow velocity in feet per minute?

191. A circular clarifier has a diameter of 70 ft. If the flow to the clarifier is 1,700 gpm, what is the surface overflow rate in gallons per minute per square foot?

192. A circular clarifier receives a flow of 3.15 MGD. If the diameter of the weir is 70 ft, what is the weir loading rate in gallons per minute per foot?

193. A circular clarifier has a diameter of 60 ft and an average water depth of 12 ft. What flow rate (MGD) corresponds to a detention time of 2 hr?

194. The flow to a sedimentation basin is 3.25 MGD. If the length of the basin is 80 ft, the width of the basin is 30 ft, and the depth of water in the basin is 14 ft, what is the mean flow velocity in feet per minute?

195. A 100 mL sample of slurry from a solids contact clarification unit is placed in a graduated cylinder and allowed to settle for 10 min. The settled sludge at the bottom of the graduated cylinder after 10 min is 26 mL. What is the percent settled sludge of the sample?

196. A raw water unit requires an alum dose of 50 mg/L, as determined by jar testing. If a "residual" 30 mg/L alkalinity must be present in the water to promote complete precipitation of the alum added, what is the total alkalinity required in milligrams per liter (1 mg/L alum reacts with 0.45 mg/L alkalinity)?

197. A sedimentation basin is 80 ft long and 30 ft wide. To maintain a surface overflow rate of 0.7 gpm/sq ft, what is the maximum flow to the basin in gallons per day?

198. The lime dose for a raw water unit has been calculated to be 14.5 mg/L. If the flow to be treated is 2,410,000 gpd, how many pounds per day lime will be required?

199. A 100 mL sample of slurry from a solids contact clarification unit is placed in a graduated cylinder. After 10 min, a total of 21 mL of sludge has settled to the bottom of the cylinder. What is the percent settled sludge of the sample?

200. A circular clarifier receives a flow of 3.24 MGD. If the diameter of the weir is 80 ft, what is the weir loading rate in gallons per minute per foot?

201. A total of 50 mg/L alkalinity is required to react with alum and ensure complete precipitation of the alum added. If the raw water has an alkalinity of 30 mg/L as bicarbonate, how many milligrams per liter alkalinity should be added to the water?

202. Given the following data, calculate the required lime dose in milligrams per liter.

Alum dose required per jar tests: 50 mg/L   1 mg/L alum reacts
                                                          w/0.45 mg/L alk.

Raw water alkalinity: 33 mg/L   1 mg/L alk. reacts
                                                           w/0.35 mg/L lime

"Residual" alk. required for precipitation: 30 mg/L

203. A total of 192 lb/day lime will be required to raise the alkalinity of the water passing through a solids contact clarifier. How many grams per minute lime does this represent (1 lb = 453.6 g)?

204. A solids contact clarification unit receives a flow of 1.5 MGD. Alum is to be used for coagulation purposes. If a lime dose of 16 mg/L is required, how many pounds per day lime is this?

205. A lime dose of 14 mg/L is required to raise the alkalinity of a raw water unit. If the flow to be treated is 2,880,000 gpd, what is the grams per minute lime dose required?

206. During an 80 hr filter run, a total of 14.2 million gallons of water is filtered. What is the average gallons per minute flow rate through the filter during this time?

207. The flow rate through a filter is 2.97 MGD. What is this flow rate expressed as gallons per minute?

208. At an average flow rate through a filter of 3,200 gpm, how long a filter run (in hours) would be required to produce 16 MG of filtered water?

209. The influent valve to a filter is closed for a 5 min period. During this time, the water level in the filter drops 12 in. If the filter is 45 ft long and 22 ft wide, what is the gallons per minute flowrate through the filter?

210. A filter is 40 ft long and 30 ft wide. To verify the flow rate through the filter, the filter influent valve is closed for a 5 min period and the water drop is measured. If the water level in the filter drops 14 in during the 5 min, what is the gallons per minute flow rate through the filter?

211. The influent valve to a filter is closed for 6 min. The water level in the filter drops 18 in during the 6 min. If the filter is 35 ft long and 18 ft wide, what is the gallons per minute flow rate through the filter?

212. A filter 20 ft long and 18 ft wide receives a flow of 1,760 gpm. What is the filtration rate in gallons per minute per square foot?

213. A filter has a surface area of 32 ft by 18 ft. If the filter receives a flow a 2,150,000 gpd, what is the filtration rate in gallons per minute per square foot?

214. A filter 38 ft long and 24 ft wide produces a total of 18.1 MG during a 71.6 hr filter run. What is the average filtration rate for this filter run?

215. A filter 33 ft long and 24 ft wide produces a total of 14.2 MG during a 71.4 hr filter run. What is the average filtration rate for this filter run?

216. A filter 38 ft long and 22 ft wide receives a flow of 3.550,000 gpd. What is the filtration rate in gallons per minute per square foot?

217. A filter is 38 ft long and 18 ft wide. During a test of filter flow rate, the influent valve to the filter is closed for 5 min. The water level drops 22 in during this period. What is the filtration rate for the filter in gallons per minute per square foot?

218. A filter is 33 ft long and 24 ft wide. During a test of flow rate, the influent valve to the filter is closed for 6 min. The water level drops 21 in during this period. What is the filtration rate for the filter gallons per minute per square foot?

219. The total water filtered between backwashes is 2.87 MG. If the filter is 20 ft by 18 ft, what is the unit filter run volume in gallons per square foot?

220. The total water filtered during a filter run is 4,180,000 gal. If the filter is 32 ft long and 20 ft wide, what is the UFRV in gallons per square foot?

221. A total of 2,980,000 gal of water is filtered during a particular filter run. If the filter is 24 ft long and 18 ft wide, what was the UFRV in gallons per square foot?

222. The average filtration rate for a filter was determined to be 3.4 gpm/sq ft. If the filter run time was 3,330 min, what was the unit filter run volume in gallons per square foot?

223. A filter ran 60.5 hr between backwashes. If the average filtration rate during that time was 2.6 gpm/sq ft, what was the UFRV in gallons per square foot?

224. A filter with a surface area of 380 sq ft has a backwash flow rate of 3,510 gpm. What is the filter backwash rate in gallons per minute per square foot?

225. A filter 18 ft long and 14 ft wide has a backwash flow rate of 3,580 gpm. What is the filter backwash rate in gallons per minute per square foot?

226. A filter has a backwash rate of 16 gpm/sq ft. What is this backwash rate expressed as inches per minute?

227. A filter 30 ft by 18 ft has a backwash flow rate of 3,650 gpm. What is the filter backwash rate in gallons per minute per square foot?

228. A filter 18 ft long and 14 ft wide has a backwash rate of 3,080 gpm. What is this backwash rate expressed as inches per minute rise?

229. A backwash flow rate of 6,650 gpm for a total backwashing period of 6 min would require how many gallons of water for backwashing?

230. For a backwash flow rate of 9,100 gpm and a total backwash time of 7 min, how many gallons of water will be required for backwashing?

231. How many gallons of water would be required to provide a backwash flow rate of 4,670 gpm for a total of 5 min?

232. A backwash flow rate of 6,750 gpm for a total of 6 min would require how many gallons of water?

233. A total of 59,200 gal of water will be required to provide a 7 min backwash of a filter. What depth of water is required in the backwash water tank to provide this backwashing capability? The tank has a diameter of 40 ft.

234. The volume of water required for backwashing has been calculated to be 62,200 gal. What is the required depth of water in the backwash water tank to provide this amount of water if the diameter of the tank is 52 ft?

235. A total of 42,300 gal of water will be required for backwashing filter. What depth of water is required in the backwash water tank to provide this much water? The diameter of the tank is 42 ft.

236. A backwash rate of 7,150 gpm is desired for a total backwash time of 7 min. What depth of water is required in the backwash water tank to provide this much water? The diameter of the tank is 40 ft.

237. A backwash rate of 8,860 gpm is desired for a total backwash time of 6 min. What depth of water is required in the backwash water tank to provide this backwashing capability? The diameter of the tank is 40 ft.

238. A filter is 42 ft and 22 ft wide. If the desired backwash rate is 19 gpm/sq ft, what backwash pumping rate (in gallons per minute) will be required?

239. The desired backwash pumping rate for a filter is 20 gpm/sq ft. If the filter is 36 ft long and 26 ft wide, what backwash pumping rate (in gallons per minute) will be required?

240. A filter is 22 ft square. If the desired backwash rate is 16 gpm/sq ft, what backwash pumping rate (in gallons per minute) will be required?

241. The desired backwash pumping rate for a filter is 24 gpm/sq ft. If the filter is 26 ft long and 22 ft wide, what backwash pumping rate (in gallons per minute) will be required?

242. A total of 17,100,000 gal of water is filtered during a filter run. If 74,200 gal of this product water are used for backwashing, what percent of the product water is used for backwashing?

243. A total of 6.10 MG of water was filtered during a filter run. If 37,200 gal of this product water were used for backwashing, what percent of the product water was used for backwashing?

244. 59,400 gal of product water are used for filter backwashing at the end of a filter run. If a total of 13,100,000 gal is filtered during the filter run, what percent of the product water is used for backwashing?

245. A total of 11,110,000 gal of water is filtered during a particular filter run. If 52,350 gal of product water are used for backwashing, what percent of the product water is used for backwashing?

246. A total 3,625 mL sample of filter media was taken for mud ball evaluation. The volume of water in the graduated cylinder rose from 600 mL to 635 mL when mud balls were placed in the cylinder. What is the percent mud ball volume of the sample?

247. Five samples of filter media are taken for mud ball evaluation. The volume of water in the graduated cylinder rose from 510 mL to 535 mL when mud balls were placed in the cylinder. What is the percent mud ball volume of the sample? The mud ball sampler has a volume of 705 mL.

248. A filter media is tested for the presence of mud balls. The mud ball sampler has a total sample volume of 705 mL. Five samples were taken from the filter. When the mud balls were placed in 520 mL of water, the water level rose of to 595 mL. What is the percent mud ball volume of the sample?

249. Five samples of media filter are taken and evaluated for the presence of mud balls. The volume of water in the graduated cylinder rises from 520 mL to 562 mL when the mud balls are placed in the water. What is the percent mud ball volume of the sample? The mud ball sample has a sample volume of 705 mL.

250. During an 80 hr filter run, a total of 11.4 million gallons of water is filtered. What is the average gallons per minute flow rate through the filter during this time?

251. A filter is 40 ft long and 25 ft wide. If the filter receives a flow of 3.56 MGD, what is the filtration rate in gallons per minute per square foot?

252. The total water filtered between backwashes is 2.88 MG. If the filter is 25 ft by 25 ft, what is the unit filter run volume in gallons per square foot?

253. At an average flow rate through a filter of 2,900 gpm, how long a filter run (in hours) would be required to produce 14.8 MG of filtered water?

254. A filter is 38 ft long and 26 ft wide. To verify the flow rate through the filter, the filter influent valve is closed for a period of 5 min and the water drop is

measured. If the water level in the filter drops 14 in during the 5 min period, what is the gallons per minute flow rate through the filter?

255. A total of 3,450,000 gal of water is filtered during a particular filter run. If the filter is 30 ft long and 25 ft wide, what is the unit filter run volume in gallons per square foot?

256. A filter 30 ft long and 20 ft wide produces a total of 13,500,000 gal during a 73.8 hr filter run. What is the average filtration rate (in gallons per minute per square foot) for this filter run?

257. A filter with a surface area of 360 sq ft has a backwash flow rate of 3,220 gpm. What is the filter backwash rate in gallons per minute per square foot?

258. A backwash flow rate of 6,350 gpm for a total backwashing period of 6 min would require how many gallons of water for backwashing?

259. The influent valve to a filter is closed for a 5 min period. During this time, the water level in the filter drops 14 in. If the filter is 30 ft long and 22 ft wide, what is the filtration rate in gallons per minute per square foot?

260. A total of 53,200 gal of water will be required to provide a 6 min backwash of a filter. What depth of water is required in the backwash water tank to provide this backwashing capability? The tank has a diameter of 45 ft.

261. The average filtration rate for a filter was determined to be 3.3 gpm/sq ft. If the filter run time was 3,620 min, what was the unit filter run volume in gallons per square foot?

262. A filter is 40 ft long and 25 ft wide. During a test of filter flow rate, the influent valve to the filter is closed for 5 min. The water level drops 20 in during this period. What is the filtration rate for the filter in gallons per minute per square foot?

263. A filter 35 ft by 25 ft has a backwash flow rate of 3,800 gpm. What is the filter backwash rate in gallons per minute per square foot?

264. How many gallons of water would be required to provide a backwash flow rate of 4,500 gpm for a total of 7 min?

265. The desired backwash pumping rate for a filter is 16 gpm/sq ft. If the filter is 30 ft long and 30 ft wide, what backwash pumping rate (in gallons per minute) will be required?

266. A filter 25 ft long and 20 ft wide has a backwash rate of 2,800 gpm. What is this backwash rate expressed as inches per minute rise?

267. A filter is 30 ft long and 25 ft wide. During a test of flow rate, the influent valve to the filter is closed for 6 min. The water level drops 18 in during this period. What is the filtration rate for the filter in gallons per minute per square foot?

268. A filter is 45 ft long and 25 ft wide. If the desired backwash rate is 18 gpm/ sq ft, what backwash pumping rate (in gallons per minute) will be required?

269. A total of 18,200,000 gal of water is filtered during a filter run. If 71,350 gal of this product water are used for backwashing, what percent of the product water is used for backwashing?

270. A total of 86,400 gal of water will be required for backwashing a filter. What depth of water is required in the backwash water tank to provide this much water? The diameter of the tank is 35 ft.

271. A total 3,480 mL sample of filter media was taken for mud ball evaluation. The volume of water in the graduated cylinder rose from 500 mL to 527 mL when the mud balls were placed in the cylinder. What is the percent mud ball volume of the sample?

272. Around 51,200 gal of product water are used for filter backwashing at the end of a filter run. If a total of 13.8 MG is filtered during the filter run, what percent of the product water is used for backwashing?

273. Five samples of filter media are taken for mud ball evaluation. The volume of water in the graduated cylinder rose from 500 mL to 571 mL when the mud balls were placed in the cylinder. What is the percent mud ball volume of the sample? The mud ball sampler has a volume of 695 mL.

274. A 20 ft by 25 ft sand filter is set to operate for 36 hr before being back-washed. Backwashing takes 15 min and uses 12 gpm/sq ft until the water is

clear. The water in the backwash storage tank drops 25 ft during backwash. Each filter run produces 3.7 MG. (a) What is the filtration rate in gallons per minute per square foot. (b) How many gallons of backwash water were used? (c) What was the percentage of product water used for backwashing? (d) If the initial tank water depth was 70 ft, how deep is the water after backwashing? (e) What is the UFRV?

275. Determine the chlorinator setting (in pounds per day) needed to treat a flow of 3.5 MGD with a chlorine dose of 1.8 mg/L.

276. A flow of 1,340,000 gpd is to receive a chlorine dose of 2.5 mL. What should the chlorinator settling be in pounds per day?

277. A pipeline 12 in in diameter and 1,200 ft long is to be treated with a chlorine dose of 52 mg/L. How many pounds of chlorine will this require?

278. A chlorinator setting is 43 lb/24 hr. If the flow being treated is 3.35 MGD, what is the chlorine dosage expressed as milligrams per liter?

279. The flow totalizer reading at 9:00 a.m. on Thursday was 18,815,108 gal. and at 9:00 a.m. on Thursday was 19,222,420 gal. If the chlorinator setting is 16 lb for this 24 hr period, what is the chlorine dosage in milligrams per liter?

280. The chlorine demand of a water process is 1.6 mg/L. If the desired chlorine residual is 0.5 mg/L, what is the desired chlorine dose in milligrams per liter?

281. The chlorine dosage for a water process is 2.9 mL. If the chlorine residual after 30 min contact time is found to be 0.7 mg/L, what is the chlorine demand expressed in milligrams per liter?

282. A flow of 3,850,000 gpd is to be disinfected with chlorine. If the chlorine demand is 2.6 mg/L and a chlorine residual of 0.8 mg/L is desired, what should be the chlorinator setting in pounds per day?

283. A chlorinator setting is increased by 6 lb/day. The chlorine residual before the increased dosage was 0.4 mg/L. After the increased dose, the chlorine residual was 0.8 mg/L. The average flow rate being treated is 1,100,000 gpd. Is the water being chlorinated beyond the breakpoint?

284. A chlorinator setting of 19 lb of chlorine per 24 hr results in a chlorine residual of 0.4 mg/L. The chlorinator setting is increased to 24 lb/24 hr. The chlorine residual increases to 0.5 mg/L at this new dosage rate. The average flow being treated is 2,100,000 gpd. On the basis of this data, is the water being chlorinated past the breakpoint?

285. A chlorine dose of 48 lb/day is required to treat a particular water unit. If calcium hypochlorite (65% available chlorine) is to be used, how many pounds per day of hypochlorite will be required?

286. A chlorine dose of 42 lb/day is required to disinfect a flow of 2,220,000 gpd. If the calcium hypochlorite to be used contains 65% available chlorine, how many pounds per day hypochlorite will be required?

287. A water flow of 928,000 gpd requires a chlorine dose of 2.7 mg/L. If calcium hypochlorite (65% available chlorine) is to be used, how many pounds per day of hypochlorite are required?

288. A total of 54 lb of hypochlorite (65% available chlorine) is used in a day. If the flow rate treated is 1,512,000 gpd, what is the chlorine dosage in milligrams per liter?

289. A flow of 3,210,000 gpd is disinfected with a calcium hypochlorite (65% available chlorine). If 49 lb of hypochlorite are used in a 24 hr period, what is the milligrams per liter chlorine dosage?

290. A total of 36 lb/day sodium hypochlorite is required for disinfection of a 1.7 MGD flow. How many gallons per day sodium hypochlorite is this?

291. A hypochlorite is used to disinfect the water pumped from a well. The hypochlorite solution contains 3% available chlorine. A chlorine dose of 2.2 mg/L is required for adequate disinfection throughout the distribution system. If the flow from the well is 245,000 gpd, how much sodium hypochlorite (in gallons per day) will be required?

292. Water from a well is disinfected by a hypochlorinator. The flow totalizer indicates that 2,330,000 gal of water were pumped during a 70-day period. The 3% sodium hypochlorite solution used to treat the well water is pumped from a 3 ft diameter storage tank. During the seven-day period, the level in the tank dropped 2 ft, 10 in. What is the chlorine dosage in milligrams per liter?

293. A hypochlorite solution (4% available chlorine) is used to disinfect a water unit. A chlorine dose of 1.8 mg/L is desired to maintain an adequate chlorine residual. If the flow being treated is 400 gpm, what hypochlorite solution flow (in gallons per day) will be required?

294. A sodium hypochlorite solution (3% available chlorine) is used to disinfect the water pumped from a well. A chlorine dose of 2.9 mg/L is required for adequate disinfection. How many gallons per day of sodium hypochlorite will be required if the flow being chlorinated is 955,000 gpd?

295. A total of 22 lb of calcium hypochlorite (65% available chlorine) is added to 60 gal of water. What is the percent chlorine (by weight) of the solution?

296. If 320 g of calcium hypochlorite are dissolved in 7 gal of water, what is the percent chlorine (by weight) of the solution (1 g = 0.0022 lb)?

297. How many pounds of dry hypochlorite (65% available chlorine) must be added to 65 gal of water to make a 3% chlorine solution?

298. A 10% liquid hypochlorite is to be used in making up a 2% hypochlorite solution. How many gallons of liquid hypochlorite should be mixed with water to produce 35 gal of a 2% hypochlorite solution?

299. How many gallons of 13% liquid hypochlorite should be mixed with water to produce 110 gal of a 1.2% hypochlorite solution?

300. If 6 gal of a 12% sodium hypochlorite solution are added to a 55 gal drum, how much water should be added to the drum to produce a 2% hypochlorite solution?

301. If 50 lb of a hypochlorite solution (11% available chlorine) are mixed with 220 lb of another hypochlorite solution (1% available chlorine), what is the percent chlorine of the solution mixture?

302. If 12 gal of a 12% hypochlorite solution are mixed with 60 gal of a 1.5% hypochlorite solution, what is the percent strength of the solution mixture?

303. If 16 gal of a 12% hypochlorite solution are added to 70 gal of 1% hypochlorite solution, what is the percent strength of the solution mixture?

304. The average calcium hypochlorite use at a plant is 44 lb/day. If the chemical inventory in stock is 1,000 lb, how many days' supply is this?

305. The average daily use of sodium hypochlorite solution at a plant is 80 gpd. A chemical feed tank has a diameter of 4 ft and contains solution to a depth of 3 ft, 8 in. How many days' supply is represented by the solution in the tank?

306. An average of 24 lb of chlorine is used each day at a plant. How many pounds of chlorine would be used in one week if the hour meter on the pump registered 150 hr of operation that week?

307. A chlorine cylinder has 91 lb of chlorine at the beginning of a week. The chlorinator setting is 12 lb per 24 hr. If the pump hour meter indicates the pump has operated a total of 111 hr during the week, how many pounds chlorine should be in the cylinder at the end of the week?

308. An average of 55 lb of chlorine is used each day at a plant. How many 150 lb chlorine cylinders will be required each month? Assume a 30-day month.

309. The average sodium hypochlorite use at a plant is 52 gpd. If the chemical feed tank is 3 ft in diameter, how many feet should the solution level in the tank drop in two days' time?

310. The chlorine demand of a water unit is 1.8 mg/L. If the desired chlorine residual is 0.9 mg/L, what is the desired chlorine dose in milligrams per liter?

311. Determine the chlorinator setting (in pounds per day) needed to treat a flow of 980,000 gpd with a chlorine dose of 2.3 mg/L.

312. A chlorine dose of 60 lb/day is required to treat a water unit. If calcium hypochlorite (65% available chlorine) is to be used, how many pounds per day of hypochlorite will be required?

313. A total of 51 lb/day sodium hypochlorite is required for disinfection of a 2.28 MGD flow. How many gallons per day sodium hypochlorite is this?

314. The chlorine dosage for a water unit is 3.1 mg/L. If the chlorine residual after 30 min contact time is found to be 0.6 mg/L, what is the chlorine demand expressed in milligrams per liter?

315. A total of 30 lb of calcium hypochlorite (65% available chlorine) is added to 66 gal of water. What is the percent chlorine (by weight) of the solution?

316. What chlorinator setting is required to treat a flow of 1,620 gpm with a chlorine dose of 2.8 mg/L?

317. A chlorine dose of 2.8 mg/L is required for adequate disinfection of a water unit. If a flow of 1.33 MGD will be treated, how many gallons per day of sodium hypochlorite will be required? The sodium hypochlorite contains 12.5% available chlorine.

318. A pipeline 8 in in diameter and 1,600 ft long is to be treated with a chlorine dose of 60 mg/L. How many pounds of chlorine will this require?

319. A chlorinator setting of 15 lb of chlorine per 24 hr results in a chlorine residual of 0.5 mg/L. The chlorinator setting is increased to 18 lb/24 hr. The chlorine residual increases to 0.6 mg/L at this new dosage rate. The average flow being treated is 2,110,000 gpd. On the basis of this data, is the water being chlorinated past the breakpoint?

320. If 70 gal of a 12% hypochlorite solution are mixed with 250 gal of a 2% hypochlorite solution, what is the percent strength of the solution mixture?

321. The average calcium hypochlorite use at a plant is 34 lb/day. If the chemical inventory in stock is 310 lb, how many days' supply is this?

322. The flow totalizer reading at 7:00 a.m. on Wednesday was 43,200,000 gal and at 7:00 a.m. on Thursday was 44,115,670 gal. If the chlorinator setting is 18 lb for this 24 hr period, what is the chlorine dosage in milligrams per liter?

323. A chlorine dose of 32 lb/day is required to disinfect a flow of 1,990,000 gpd. If the calcium hypochlorite to be used contains 60% available chlorine, how many pounds per day hypochlorite will be required?

324. Water from a well is disinfected by a hypochlorinator. The flow totalizer indicates that 2,666,000 gal of water were pumped during a seven-day period. The 2% sodium hypochlorite solution used to treat the well water is pumped from a 4 ft diameter storage tank. During the seven-day period, the level in the tank dropped 3 ft, 4 in. What is the chlorine dosage in milligrams per liter?

325. A flow of 3,350,000 gpd is to be disinfected with chlorine. If the chlorine demand is 2.5 mg/L and a chlorine residual of 0.5 mg/L is desired, what should be the chlorinator setting in pounds per day?

326. If 12 gal of a 12% hypochlorite solution are mixed with 50 gal of a 1% hypochlorite solution, what is the percent strength of the solution mixture?

327. A total of 72 lb of hypochlorite (65% available chlorine) is used in a day. If the flow rate treated is 1,880,000 gpd, what is the chlorine dosage in milligrams per liter?

328. A hypochlorite solution (3% available chlorine) is used to disinfect a water unit. A chlorine dose of 2.6 mg/L is desired to maintain adequate chlorine residual. If the flow being treated is 400 gpm, what hypochlorite solution flow (in gallons per day) will be required?

329. The average daily use of sodium hypochlorite at a plant is 92 gpd. The chemical feed tank has a diameter of 3 ft and contains solution to a depth of 4 ft, 1 in. How many days' supply are represented by the solution in the tank?

330. How many pounds of dry hypochlorite (65% available chlorine) must be added to 80 gal of water to make a 2% chlorine solution?

331. An average of 32 lb of chlorine is used each day at a plant. How many pounds of chlorine would be used in one week if the hour meter on the pump registers 140 hr of operation that week?

332. An average of 50 lb of chlorine is used each day at a plant. How many 150 lb chlorine cylinders will be required each month? Assume a 30-day month.

333. Express 2.6% concentration in terms of milligrams per liter concentration.

334. Convert 6,700 mg/L to percent.

335. Express 29% concentration in terms of milligrams per liter.

336. Express 22 lb/MG concentration as milligrams per liter.

337. Convert 1.6 mg/L to pounds per million gallons.

338. Express 25 lb/MG concentration as milligrams per liter concentration.

339. Given the atomic weights for H = 1.008, Si = 28.06, and F = 19.00, calculate the percent fluoride ion present in hydrofluosilicic acid, H_2SiF_6.

340. Given the atom weights for Na = 22,997 and F = 19.00, calculate the percent fluoride ion present in sodium fluoride, NaF.

341. A fluoride dosage of 1.6 mg/L is desired. The flow to be treated is 980,000 gpd. How many pounds per day dry sodium silicofluoride (Na_2SiF_6) will be required if the commercial purity of Na_2SiF_6 will be required if the commercial purity of the Na_2SiF_6 is 98% and the percent of fluoride ion in the compound is 60.6%? Assume the raw water contains no fluoride.

342. A fluoride dosage of 1.4 mg/L is desired. How many pounds per day dry sodium silicofluoride (Na_2SiF_6) will be required if the flow to be treated is 1.78 MGD? The commercial purity of the sodium silicofluoride is 98%, and the percent of fluoride ion in Na_2SiF_6 is 60.6%. Assume the water to be treated contains no fluoride.

343. A flow of 2,880,000 gpd is to be treated with sodium silicofluoride, Na_2SiF_6. The raw water contains no fluoride. If the desired fluoride concentration in the water is 1.4 mg/L, what should be the chemical feed rate (in pounds per day)? The manufacturer's data indicates that each pound of Na_2SiF_6 contains 0.8 lb of fluoride ion. Assume the raw water contains no fluoride.

344. A flow of 3.08 MGD is to be treated with sodium fluoride, NaF. The raw water contains no fluoride, and the desired fluoride concentration in the finished water is 1.1 mg/L. What should be the chemical feed rate in pounds per day? Each pound of NaF contains 0.45 lb of fluoride ion.

345. A flow of 810,000 gpd is to be treated with sodium fluoride, NaF. The raw water contains 0.08 mg/L fluoride, and the desired fluoride level in the finished water is 1.2 mg/L. What should be the chemical feed rate in pounds per day? NaF contains 0.45 lb of fluoride ion.

346. If 9 lb of 98% pure sodium fluoride (NaF) are mixed with 55 gal of water, what is the percent strength of the solution.

347. If 20 lb of 100% pure sodium fluoride, NaF, are dissolved in 80 gal of water, what is the percent strength of the solution?

348. How many pounds of 98% pure sodium fluoride must be added to 220 gal of water to make a 1.4% solution of sodium fluoride?

349. If 11 lb of 98% pure sodium fluoride are mixed with 60 gal water, what is the percent strength of the solution?

350. How many pounds of 98% pure sodium fluoride must be added to 160 gal of water to make a 3% solution of sodium fluoride?

351. A flow of 4.23 MGD is to be treated with a 24% solution of hydrofluosilicic acid. The acid has specific gravity of 1.2. If the desired fluoride level in the finished water is 1.2 mg/L, what should be the solution feed rate (in gallons per day)? The raw water contains no fluoride. The percent fluoride ion content of H_2SiF_6 is 80%.

352. A flow of 3.1 MGD non-fluoridated water is to be treated with a 22% solution of hydrofluosilicic acid (H_2SiF_6). The desired fluoride concentration is 1.2 mg/L. What should be the solution feed rate (in gallons per day)? The hydrofluosilicic acid weighs 9.7 lb/gal. The percent fluoride ion content of H_2SiF_6 is 80%.

353. A flow of 910,000 gpd is to be treated with a 2.2% solution of sodium fluoride, NaF. If the desired fluoride ion concentration is 1.8 mg/L, what should be the sodium fluoride feed rate (in gallons per day)? Sodium fluoride has fluoride ion content of 46.10%. The water to be treated contains 0.09 mg/L fluoride ion. Assume the solution density is 8.34 lb/day.

354. A flow of 1,520,000 gpd non-fluoridated water is to be treated with a 2.4% solution of sodium fluoride, NaF. The desired fluoride level in the finished water is 1.6 mg/L. What should be the sodium fluoride solution feed rate (in gallons per day)? Sodium fluoride has a fluoride ion content of 45.25%. Assume the solution density is 8.34 lb/gal.

355. The desired solution feed rate has been determined to be 80 gpd. What is this feed rate expressed as milliliters per minute?

356. A flow of 2.78 MGD is to be treated with a 25% solution of hydrofluosilicic acid, H_2SiF_6 (fluoride content of 80%). The raw water contains no fluoride, and the desired fluoride concentration is 1.0 mg/L. The hydrofluosilicic acid weighs 9.8 lb/gal. What should be the milliliters per minute solution feed rate?

357. A total of 40 lb/day of 98% pure sodium silicofluoride (Na_2SiF_6) was added to a flow of 1,520,000 gpd. The percent fluoride ion content of Na_2SiF_6 is 61%. What was the concentration of fluoride ion in the treated water?

358. A flow of 330,000 gpd is treated with 6 lb/day sodium fluoride, NaF. The commercial purity of the sodium fluoride is 98%, and the fluoride ion content of NaF is 45.25%. Under these conditions, what is the fluoride ion dosage in milligrams per liter?

359. A flow of 3.85 MGD non-fluoridated water is treated with a 20% solution of hydrofluosilicic acid, H_2SiF_6. If the solution feed rate is 32 gpd, what is the calculated fluoride ion concentration of the treated water? Assume the acid weighs 9.8 lb/gal and the percent fluoride ion in H_2SiF_6 is 80%.

360. A flow of 1,920,000 gpd non-fluoridated water is treated with 11% solution of hydrofluosilicic acid, H_2SiF_6. If the solution feed rate is 28 gpd, what is the calculated fluoride ion concentration of the finished water? The acid weights 9.10 lb/gal, and the percent fluoride ion in H_2SiF_6 is 80%.

361. A flow of 2,730,000 gpd non-fluoridated water is to be treated with a 3% saturated solution of sodium fluoride, NaF. If the solution feed rate is 110 gpd, what is the calculated fluoride ion level in the finished water? Assume the solution weighs 8.34 lb/gal. The percent fluoride ion in NaF is 45.25%.

362. A tank contains 600 lb of 15% hydrofluosilicic acid, H_2SiF_6. If 2,600 lb of 25% hydrofluosilicic acid are added to the tank, what is the percent strength of the solution mixture?

363. If 900 lb of 25% hydrofluosilicic acid were added to a tank containing 300 lb of 15% hydrofluosilicic acid, what would be the percent strength of the solution mixture?

364. A tank containing 16% hydrofluosilicic acid (H_2SiF_6) contains 400 gal. If a tanker truck delivers 2,200 gal of 22% hydrofluosilicic acid to be added to the tank, what is the percent strength of the solution mixture? Assume the 22% solution weighs 9.10 lb/gal and the 16% solution weighs 9.4 lb/gal.

365. A tank contains 325 gal of an 11% hydrofluosilicic acid. If 1,100 gal of a 20% hydrofluosilicic acid are added to the tank, what is the percent strength of the solution mixture? Assume the 11% acid weighs 9.06 lb/gal and the 20% acid weighs 9.8 lb/gal.

366. A tank contains 220 gal of a 10 % hydrofluosilicic acid with a specific gravity of 1.075. If the total of 1,600 gal of 15% hydrofluosilicic acid is added to the tank, what is the percent strength of the solution mixture? Assume the 15% acid solution weighs 9.5 lb/gal.

367. Express 2.9% concentration in terms of milligrams per liter concentration.

368. Calculate the percent fluoride ion present in hydrofluosilicic acid, H_2SiF_6. The atomic weights are as follows: H = 1.008; Si = 28.06; F = 19.00.

369. Express 27 lb/MG concentration as milligrams per liter concentration.

370. A fluoride ion dosage of 1.6 mg/L is desired. The non-fluoridated flow to be treated is 2,111,000 gpd. How many pounds per day dry 98% pure sodium silicofluoride will be required if the percent fluoride ion in the compound is 61.2%.

371. Calculate the percent fluoride ion present in sodium fluoride, NaF. The atomic weights are as follows: Na = 22.997 and F = 19.00.

372. If 80 lb of 98% pure sodium fluoride (NaF) are mixed with 600 gal of water, what is the percent strength of the solution?

373. Convert 28,000 mg/L to percent.

374. The desired solution feed rate has been determined to be 80 gpd. What is this feed rate expressed in milliliters per minute?

375. A fluoride dosage of 1.5 mg/L is desired. How many pounds per day of 98% pure dry sodium fluoride (NaF) will be required if the flow to be treated is 2.45 MGD? The percent fluoride ion in NaF is 45.25%.

376. How many pounds of sodium fluoride (98% pure) must be added to 600 gal of water to make a 3% solution of sodium fluoride?

377. A flow of 4.11 MGD non-fluoridated water is to be treated with a 21% solution of hydrofluosilicic acid. The acid has a specific gravity of 1.3. If the desired fluoride level in the finished water is 1.4 mg/L, what should be the solution feed rate (in gallons per day)? The percent fluoride ion content of the acid is 80%.

378. If 30 lb of 98% pure sodium fluoride are mixed with 140 gal water, what is the percent strength of the solution?

379. A flow of 1,880,000 gpd is to be treated with sodium fluoride, NaF (contains 0.44 lb of fluoride ion). The raw water contains 0.09 mg/L, and the desired fluoride level in the finished water is 1.4 mg/L. What should be the chemical feed rate in pounds per day?

380. A flow of 2.8 MGD non-fluoridated water is to be treated with a 20% solution of hydrofluosilicic acid. The desired fluoride concentration is 1.3 mg/L. What should be the solution feed rate (in gallons per day)? The

hydrofluosilicic acid weighs 9.8 lb/gal. The percent fluoride ion content of acid is 80%.

381. A tank contains 500 lb of 15% hydrofluosilicic acid. If 1,600 lb of 20% hydrofluosilicic acid are added to the tank, what is the percent strength of the solution mixture?

382. A total of 41 lb/day of sodium silicofluoride was added to a flow of 1,100,000 gpd. The commercial purity of sodium silicofluoride was 98%, and the percent fluoride ion content of acid is 61%. What was the concentration of fluoride in the treated water?

383. A flow of 1,400 gpm non-fluoridated raw water is treated with an 11% solution of hydrofluosilicic acid. If the solution feed rate is 40 gpd, what is the calculated fluoride ion concentration of the finished water? The acid weighs 9.14 lb/gal, and the percent fluoride ion in the acid is 80%.

384. A tank contains 235 gal of 10% hydrofluosilicic acid. If 600 gal of a 20% hydrofluosilicic acid are added to the tank, what is the percent strength of the solution mixture? Assume the 10% acid weighs 9.14 lb/gal and the 20% acid weighs 9.8 lb/gal.

385. A flow of 2.88 MGD is to be treated with a 20% solution of hydrofluosilicic acid. The raw water contains no fluoride, and the desired fluoride concentration is 1.1 mg/L. The acid weighs 9.8 lb/gal. What should be the milliliters per minute solution feed rate? The percent fluoride content of acid is 80%.

386. A tank contains 131 gal of a 9% hydrofluosilicic acid with a specific gravity of 1.115. If a total of 900 gal of 15% hydrofluosilicic acid is added to the tank, what is the percent strength of the solution mixture? Assume the 15% acid solution weighs 9.4 lb/gal.

387. A flow of 2,900,000 gpd is to be treated with a 5% saturated solution of sodium fluoride, NaF. If the solution feed rate is 120 gpd, what is the calculated fluoride ion level in the finished water? Assume the solution weighs 8.34 lb/gal. The percent fluoride ion in the acid is 45.25%. The raw water contains 0.2 mg/L fluoride.

388. The calcium content of a water sample is 39 mg/L. What is this calcium hardness expressed as $CaCO_3$? The equivalent weight of calcium is 20.04, and the equivalent weight of $CaCO_3$ is 50.045.

389. The magnesium content of a water unit is 33 mg/L. What is this magnesium hardness expressed as $CaCO_3$? The equivalent weight of magnesium is 12.16, and the equivalent weight of $CaCO_3$ is 50.045.

390. A water unit contains 18 mg/L calcium. What is this calcium hardness expressed as $CaCO_3$? The equivalent weight of calcium is 20.04, and $CaCO_3$ is 50.045.

391. A water unit has a calcium concentration of 75 mg/L as $CaCO_3$ and a magnesium concentration of 91 mg/L as $CaCO_3$. What is the total hardness (as $CaCO_3$) of the sample?

392. Determine the total hardness as $CaCO_3$ of a water unit that has calcium content of 30 mg/L and magnesium content of 10 mg/L. The equivalent weight of calcium is 20.04, of magnesium is 12.16, and of $CaCO_3$ is 50.045.

393. Determine the total hardness as $CaCO_3$ of a water unit that has calcium content of 21 mg/L and magnesium content of 15 mg/L. The equivalent weight of calcium is 20.04, of magnesium is 12.16, and of $CaCO_3$ is 50.045.

394. A sample of water contains 125 mg/L alkalinity as $CaCO_3$. If the total hardness of the water is 121 mg/L as $CaCO_3$, what is the carbonate and noncarbonate hardness in milligrams per liter as $CaCO_3$?

395. The alkalinity of a water unit is 105 mg/L as $CaCO_3$. If the total hardness of the water is 122 mg/L as $CaCO_3$, what is the carbonate and noncarbonate hardness in milligrams per liter as $CaCO_3$?

396. A water unit has an alkalinity of 91 mg/L as $CaCO_3$ and a total hardness of 116 mg/L. What is the carbonate and noncarbonate hardness of the water?

397. A water sample contains 112 mg/L alkalinity as $CaCO_3$ and 99 mg/L total hardness as $CaCO_3$. What is the carbonate and noncarbonate hardness of this water?

398. The alkalinity of a water unit is 103 mg/L as $CaCO_3$. If the total hardness of the water is 121 mg/L as $CaCO_3$, what is the carbonate and noncarbonate hardness in milligrams per liter as $CaCO_3$?

399. A 100 mL water sample is tested for phenolphthalein alkalinity. If 2 mL titrant is used to pH 8.3 and the sulfuric acid solution has a normality of 0.02N, what is the phenolphthalein alkalinity of the water in milligrams per liter as $CaCO_3$?

400. A 100 mL water sample is tested for phenolphthalein alkalinity. If 1.4 titrant is used to pH 8.3 and the normality of the sulfuric acid solution is 0.02N, what is the phenolphthalein alkalinity of the water in milligrams per liter as $CaCO_3$?

401. A 100 mL sample of water is tested for alkalinity. The normality of the sulfuric acid used for titrating is 0.02N. If 0.3 mL titrant is used to pH 8.3 and 6.7 titrant is used to pH 4.4, what is the phenolphthalein and total alkalinity of the sample?

402. A 100 mL sample of water is tested for phenolphthalein and total alkalinity. A total of 0 mL titrant is used to pH 8.3, and a total of 6.9 mL titrant is used to titrate to pH 4.4. The normality of the acid used for titrating is 0.02N. What is the phenolphthalein and total alkalinity of the sample in milligrams per liter as $CaCO_3$?

403. A 100 mL sample of water is tested for alkalinity. The normality of the sulfuric acid used for titrating is 0.02N. If 0.5 mL titrant is used to pH 8.3 and 5.7 mL titrant is used to pH 4.6, what is the phenolphthalein and total alkalinity of the sample?

Alkalinity, Mg/L as $CaCO_3$

Results of Titration	Bicarbonate Alk	Carbonate Alk	Hydroxide Alk
P = O	T	O	O
P < ½ T	T − 2P	2P	O
P = ½ T	O	2P	O
P > ½ T	O	2T − 2P	2P − 2T

Alkalinity, Mg/L as CaCO₃

Results of Titration	Bicarbonate Alk	Carbonate Alk	Hydroxide Alk
P = T	O	O	T

Where:
P = phenolphthalein alkalinity
T = total alkalinity

404. A water sample is tested for phenolphthalein and total alkalinity. If the phenolphthalein alkalinity is 8 mg/L as CaCO₃ and the total alkalinity is 51 mg/L as CaCO₃, what is the bicarbonate, carbonate, and hydroxide alkalinity of the water?

405. A water sample is found to have a phenolphthalein alkalinity of 0 mg/L and a total alkalinity of 67 mg/L. What is the bicarbonate, carbonate, and hydroxide alkalinity of the water?

406. The phenolphthalein alkalinity of a water sample is 12 mg/L as CaCO₃, and the total alkalinity is 23 mg/L as CaCO₃. What is the bicarbonate, carbonate, and hydroxide alkalinity of the water?

407. Alkalinity titrations on a 100 mL water sample resulted as follows: 1.3 mL titrant used to pH 8.3; 5.3 mL total titrant used to pH 4.6. The normality of the sulfuric acid was 0.02N. What is the phenolphthalein, total, bicarbonate, carbonate, and hydroxide alkalinity of the water?

408. Alkalinity titrations on a 100 mL water sample resulted as follows: 1.5 mL titrant used to pH 8.3; 2.9 mL total titrant used to pH 4.5. The normality of

the sulfuric acid was 0.02N. What is the phenolphthalein, total, bicarbonate, carbonate, and hydroxide alkalinity of the water?

409. Assuming 15% excess lime, a water sample has a carbon dioxide content of 8 mg/L as CO_2, total alkalinity of 130 mg/L as $CaCO_3$, and magnesium content of 22 mg/L as MG^{+2}. Approximately how much quicklime (CaO) (90% purity) will be required for softening?

410. Assuming 15% excess lime, the characteristics of a water unit are as follows: 5 mg/L CO_2, 164 mg/L total alkalinity as $CaCO_3$, and 17 mg/L magnesium as Mg^{+2}. What is the estimated hydrated lime $(Ca(OH)_2)$ (90% pure) dosage in milligrams per liter required for softening?

411. Assuming 15% excess lime, a water sample has the following characteristics: 6 mg/L CO_2, 110 mg/L total alkalinity as $CaCO_3$, and 12 mg/L magnesium as Mg^{+2}. What is the estimated hydrated lime $(Ca(OH)_2)$ (90% purity) dosage in milligrams per liter required for softening?

412. A water sample has a carbon dioxide content of 9 mg/L as CO_2, total alkalinity of 180 mg/L as $CaCO_3$, and magnesium content of 18 mg/L as MG^{+2}. Approximately how much quicklime (CaO) (90% purity) will be required for softening?

413. A water unit has a total hardness of 260 mg/L as $CaCO_3$ and a total alkalinity of 169 mg/L. What soda ash dosage (in milligrams per liter) will be required to remove the noncarbonate hardness?

414. The alkalinity of a water unit is 111 mg/L as $CaCO_3$, and the total hardness is 240 mg/L as $CaCO_3$. What soda ash dosage (in milligrams per liter) is required to remove the noncarbonate hardness?

415. A water sample has a total hardness of 264 mg/L as $CaCO_3$ and a total alkalinity of 170 mg/L. What soda ash dosage (in milligrams per liter) will be required to remove the noncarbonate hardness?

416. Calculate the soda ash required (in milligrams per liter) to soften a water unit if the water has a total hardness of 228 mg/L and a total alkalinity of 108 mg/L.

417. The A, B, C, and D factors of the excess lime equation have been calculated as follows: A = 8 mg/L; B = 130 mg/L; C = 0; D = 66 mg/L. If the residual magnesium is 4 mg/L, what is the carbon dioxide dosage (in milligrams per liter) required for recarbonation?

418. The A, B, C, and D factors of the excess lime equation have been calculated as: A = 8 mg/L; B = 90 mg/L; C =7; D =108 mg/L. If the residual magnesium is 3 mg/L, what carbon dioxide dosage would be required for recarbonation?

419. The A, B, C, and D factors of the excess lime equation were determined to be as follows: A = 7 mg/L; B = 109 mg/L; C =3; D =52 mg/L. The magnesium residual is 5 mg/L. What is the carbon dioxide dosage required for recarbonation?

420. The A, B, C, and D factors of the excess lime equation have determined to be: A = 6 mg/L; B = 112 mg/L; C = 6; D = 45 mg/L. If the residual magnesium is 4 mg/L, what carbon dioxide dosage would be required for recarbonation?

421. Jar tests indicated that the optimum lime dosage is 200 mg/L. If the flow to be treated is 2.47 MGD, what should be the chemical feeder setting in pounds per day?

422. The optimum lime dosage for a water unit has been determined to be 180 mg/L. If the flow to be treated is 3,120,000 gpd, what should be the chemical feeder setting in pounds per day and pounds per minute?

423. A soda ash dosage of 60 mg/L is required to remove noncarbonate hardness. What should be the pounds per hour chemical feeder setting if the flow rate to be treated is 4.20 MGD?

424. What should the chemical feeder setting be, in pounds per day and pounds per minute, if the optimum lime dosage has been determined to be 130 mg/L and the flow to be treated is 1,850,000 gpd?

425. A total of 40 mg/L soda ash is required to remove noncarbonate hardness from a water process. What should be the chemical feeder setting in pounds per hour and pounds per minute if the flow to be treated is 3,110,000 gpd?

426. The total hardness of a water unit is 211 mg/L. What is this hardness expressed as grains per gallon?

427. The total hardness of a water sample is 12.3 mg/L. What is this concentration expressed as milligrams per liter?

428. The total hardness of a water unit is reported as 240 mg/L. What is the hardness expressed as grains per gallon?

429. A hardness of 14 gpg is equivalent to how many milligrams per liter?

430. The hardness removal capacity of an ion exchange resin is 25,000 grains/ cu ft. If the softener contains a total of 105 cu ft of resin, what is the exchange capacity of the softener (in grains)?

431. An ion exchange water softener has a diameter of 6 ft. The depth of resin is 4.2 ft. If the resin has a removal capacity of 25 kilograins/cu ft, what is the exchange capacity of the softener (in grains)?

432. The hardness removal capacity of an exchange resin is 20 kilograins/cu ft. If the softener contains a total of 260 cu ft of resin, what is the exchange capacity of the softener (in grains)?

433. An ion exchange water softener has a diameter of 8 ft. The depth of resin is 5 ft. If the resin has a removal capacity of 22 kilograins/cu ft, what is the exchange capacity of the softener (in grains)?

434. An ion exchange softener has an exchange capacity of 2,210,000 grains. If the hardness of the water to be treated is 18.1 gpg, how many gallons of water can be treated before regeneration of the resin is required?

435. The exchange capacity of an ion exchange softener is 4,200,000 grains. If the hardness of the water to be treated is 16.0 grains/gal, how many gallons of water can be treated before regeneration of the resin is required?

436. An ion exchange softener has an exchange capacity of 3,650,000 grains. If the hardness of the water is 270 mg/L, how many gallons of water can be treated before regeneration of the resin is required?

437. The hardness removal capacity of an ion exchange resin is 21 kilograins/ cu ft. The softener contains a total of 165 cu ft of resin. If the water to be treated contains 14.6 gpg hardness, how many gallons of water can be treated before regeneration of the resin is required?

438. The hardness removal capacity of an ion exchange resin is 22,000 grains/ cu ft. The softener has a diameter of 3 ft and a depth of resin of 2.6 ft. If the water to be treated contains 12.8 gpg hardness, how many gallons of water can be treated before regeneration of the resin is required?

439. An ion exchange softener can treat a total of 575,000 gal of water before regeneration is required. If the flow rate treated is 25,200 gph, how many hours of operation are there before regeneration is required?

440. An ion exchange softener can treat a total of 766,000 gal of water before regeneration of the resin is required. If the water is to be treated at a rate of 26,000 gph, how many hours of operation are there until regeneration is required?

441. A total of 348,000 gal of water can be treated by an ion exchange water softener before regeneration of the resin is required. If the flow rate to be treated is 230 gpm, what is the operating time (in hrs) until regeneration of the resin will be required?

442. The exchange capacity of an ion exchange softener is 3,120,000 grains. The water to be treated contains 14 gpg total hardness. If the flow rate to be treated is 200 gpm, how many hours of operation are there until regeneration of the resin will be required?

443. The exchange capacity of an ion exchange softener is 3,820,000 grains. The water to be treated contains 11.6 gpg total hardness. If the flow rate to be treated is 290,000 gpd, how many hours of operation are there until regeneration of the resin will be required?

444. An ion exchange softener will remove 2,300 kilograins hardness from the water until the resin must be regenerated. If 0.5 lb salt are required for each kilograin removed, how may pounds of salt will be required for preparing the brine to be used in resin regeneration?

445. A total of 1,330 kilograins hardness is removed by an ion exchange softener before the resin must be regenerated. If 0.4 lb salt is required for each kilograin removed, how many pounds of salt will be required for preparing the brine to be used in resin regeneration?

Salt Solutions

% NaCl	Lb NaCl/Gal	Lb NaCl/Cu Ft
10	0.874	6.69
11	0.990	7.41
12	1.09	8.14
13	1.19	8.83
14	1.29	9.63
15	1.39	10.4

446. Around 410 lb salt are required in making up a brine solution for regeneration. If the brine solution is to be a 13% solution of salt, how many gallons of brine will be required for regeneration of the softener? (Use the salt solution table to determine the pounds salt per gallon brine for a 13% solution.)

447. A total of 420 lb salt will be required to regenerate an ion exchange softener. If the brine solution is to be a 14% brine solution, how many gallons brine will be required? (Use the salt solutions table to determine the pounds salt per gallon brine for a 14% brine solution.)

448. An ion exchange softener removes 1,310 kilograins hardness from the water before the resin must be regenerated, and 0.5 lb salt is required for each kilograin hardness removed. If the brine solution is to be a 12% brine solution, how many gallons of brine will be required for regeneration of the softener? (Use the salt solutions table to determine the pounds salt per gallon brine for a 12% brine solution.)

449. The calcium content of a water sample is 44 mg/L. What is this calcium hardness expressed as $CaCO_3$? The equivalent weight of calcium is 20.04, and the equivalent weight of $CaCO_3$ is 50.045.

450. A 100 mL sample of water is tested for phenolphthalein alkalinity. If 1.8 mL titrant is used to pH 8.3 and the normality of the sulfuric acid solution is 0.02N, what is the phenolphthalein alkalinity of the water in milligrams per liter as $CaCO_3$?

451. The magnesium content of a water sample is 31 mg/L. What is this magnesium hardness expressed as $CaCO_3$? The equivalent weight of magnesium is 12.16, and the equivalent weight of $CaCO_3$ is 50.045.

452. Around 24 mg of magnesium are equal to how many milliequivalents of calcium?

453. The characteristics of a water unit are as follows: 8 mg/L CO_2, 118 mg/L total alkalinity as $CaCO_3$, 12 mg/L magnesium as Mg^{+2}. What is the estimated hydrated lime $(Ca(OH)_2)$ (90% pure) dosage in milligrams per liter required for softening?

454. Determine the total hardness of $CaCO_3$ of a water unit that has calcium content of 31 mg/L and magnesium content of 11 mg/L. The equivalent weight of calcium is 20.04, of magnesium is 12.16, and of $CaCO_3$ is 50.045.

455. A sample of water contains 112 mg/L alkalinity as $CaCO_3$ and 101 mg/L total hardness of $CaCO_3$. What is the carbonate and non-carbonate hardness of this water?

456. A water sample has a carbon dioxide content of 5 mg/L as CO_2, total alkalinity of 156 mg/L as $CaCO_3$, and magnesium content of 11 mg/L as MG^{+2}.

Approximately how much quicklime (CaO) (90% purity) will be required for softening? (Assume 15% excess lime.)

9.8 ADDITIONAL PRACTICE PROBLEMS

9.8.1 TANK VOLUME CALCULATIONS

1. The diameter of a tank is 70 ft. If the water depth is 25 ft, what is the volume of water in the tank in gallons?

2. A tank is 60 ft in length, 20 ft wide, and 10 ft deep. Calculate the cubic feet volume of the tank.

3. A tank 20 ft wide and 60 ft long is filled with water to a depth of 12 ft. What is the volume of the water in the tank (in gallons)?

4. What is the volume of water in a tank, in gallons, if the tank is 20 ft wide, 40 ft long, and contains water to a depth of 12 ft?

5. A tank has a diameter of 60 ft and a depth of 12 ft. Calculate the volume of water in the tank in gallons.

6. What is the volume of water in a tank, in gallons, if the tank is 20 ft wide, 50 ft long, and contains water to a depth of 16 ft?

9.9   CHANNEL AND PIPELINE CAPACITY CALCULATIONS

7. A rectangular channel is 340 ft in length, 4 ft in depth, and 6 ft wide. What is the volume of water in cubic feet?

8. A replacement section of 10 inch pipe is to be sandblasted before it is put into service. If the length of pipeline is 1,600 ft, how many gallons of water will be needed to fill the pipeline?

9. A trapezoidal channel is 800 ft in length, 10 ft wide at the top, 5 ft wide at the bottom, with a distance of 4 ft from top edge to bottom along the sides. Calculate the gallon volume.

10. A section of 8 inch diameter pipeline is to be filled with treated water for distribution. If the pipeline is 2,250 ft in length, how many gallons of water will be distributed?

11. A channel is 1,200 ft in length, carries water 4 ft in depth, and is 5 ft wide. What is the volume of water in gallons?

9.10   MISCELLANEOUS VOLUME CALCULATIONS

12. A pipe trench is to be excavated that is 4 ft wide, 4 ft deep, and 1,200 ft long. What is the volume of the trench in cubic yards?

13. A trench is to be excavated that is 3 ft wide, 4 ft deep, and 500 yd long. What is the cubic yard volume of the trench?

14. A trench is 300 yd long, 3 ft wide, and 3 ft deep. What is the cubic feet volume of the trench?

15. A rectangular trench is 700 ft long, 6.5 ft wide, and 3.5 ft deep. What is the cubic feet volume of the trench?

16. The diameter of a tank is 90 ft. If the water depth in the tank is 25 ft, what is the volume of water in the tank in gallons?

17. A tank is 80 ft long, 20 ft wide, and 16 ft deep. What is the cubic feet volume of the tank?

18. How many gallons of water will it take to fill an 8 in diameter pipe that is 4,000 ft in length?

19. A trench is 400 yd long, 3 ft wide, and 3 ft deep. What is the cubic feet volume of the trench?

20. A trench is to be excavated. If the trench is 3 ft wide, 4 ft deep, and 1,200 ft long, what is the cubic yard volume of the trench?

21. A tank is 30 ft wide and 80 ft long. If the tank contains water to a depth of 12 ft, how many gallons of water are in the tank?

22. What is the volume of water (in gallons) contained in a 3,000 ft section of channel if the channel is 8 ft wide and the water depth is 3.5 ft?

23. A tank has a diameter of 70 ft and a depth of 19 ft. What is the volume of water in the tank in gallons?

24. If a tank is 25 ft in diameter and 30 ft deep, how many gallons of water will it hold?

9.11 FLOW, VELOCITY, AND CONVERSION CALCULATIONS

25. A channel 44 in wide has water flowing to a depth of 2.4 ft. If the velocity of the water is 2.5 fps, what is the cubic feet per minute flow in the channel?

26. A tank is 20 ft long and 12 ft wide. With the discharge valve closed, the influent to the tank causes the water level to rise 0.8 ft in 1 min. What is the gallons per minute flow to the tank?

27. A trapezoidal channel is 4 ft wide at the bottom and 6 ft wide at the water surface. The water depth is 40 in. If the flow velocity through the channel is 130 ft/min, what is the cubic feet per minute flow rate through the channel?

28. An 8 in diameter pipeline has water flowing at a velocity of 2.4 fps. What is the gallons per minute flow rate through the pipeline? Assume the pipe is flowing full.

29. A pump discharges into a 3 ft diameter container. If the water level in the container rises 28 in in 30 sec, what is the gallons per minute flow into the container?

30. A 10 in diameter pipeline has water flowing at a velocity of 3.1 fps. What is the gallons per minute flow rate through the pipeline if the water is flowing at a depth of 5 in?

31. A channel has a rectangular cross section. The channel is 6 ft wide, with water flowing to a depth of 2.6 ft. If the flow rate through the channel is 14,200 gpm, what is the velocity of the water in the channel (in foot per second)?

32. An 8 in diameter pipe flowing full delivers 584 gpm. What is the velocity of flow in the pipeline (in foot per second)?

33. A special dye is used to estimate the velocity of flow in an interceptor line. The dye is injected into the water at one pumping station, and the travel time to the first manhole 550 ft away is noted. The dye first appears at the downstream manhole in 195 sec. The dye continues to be visible until the total elapsed time is 221 sec. What is the feet per second velocity of flow through the pipeline?

34. The velocity in a 10 in diameter pipeline is 2.4 ft/sec. If the 10 in pipeline flows into an 18 in diameter pipeline, what is the velocity in the 8 in pipeline in feet per second?

35. A float travels 500 ft in a channel in 1 min, 32 sec. What is the estimated velocity in the channel (in foot per second)?

36. The velocity in an 8 in diameter pipe is 3.2 ft/sec. If the flow then travels through a 10 in diameter section of pipeline, what is the feet per second velocity in the 10 in pipeline?

9.12 AVERAGE FLOW RATES

37. The following flows were recorded for the week:

    Monday: 4.8 MGD
    Tuesday: 5.1 MGD
    Wednesday: 5.2 MGD
    Thursday: 5.4 MGD
    Friday: 4.8 MGD
    Saturday: 5.2 MGD
    Sunday: 4.8 MGD

    What was the average daily flow rate for the week?

38. The totalizer reading the month of September was 121.4 MG. What was the average daily flow (ADF) for the month of September?

9.13 FLOW CONVERSIONS

39. Convert 0.165 MGD to gallons per minute.

40. The total flow for one day at a plant was 3,335,000 gal. What was the average gallons per minute flow for that day?

41. Express a flow of 8 cfs in terms of gallons per minute.

42. What is 35 gps expressed as gallons per day?

43. Convert a flow of 4,570,000 gpd to cubic feet per minute.

44. What is 6.6 MGD expressed as cubic feet per second?

45. Express 445,875 cfd as gallons per minute.

46. Convert 2,450 gpm to gallons per day.

9.14 GENERAL FLOW AND VELOCITY CALCULATIONS

47. A channel has a rectangular cross section. The channel is 6 ft wide, with water flowing to a depth of 2.5 ft. If the flow rate through the channel is 14,800 gpm, what is the velocity of the water in the channel (in foot per second)?

48. A channel 55 in wide has water flowing to a depth of 3.4 ft. If the velocity of the water is 3.6 fps, what is the cubic feet per minute flow in the channel?

49. The following flows were recorded for the months of June, July, and August: June, 102.4 MG; July, 126.8 MG; August, 144.4 MG. What was the average daily flow for this three-month period?

50. A tank is 12 ft by 12 ft. With the discharge valve closed, the influent to the tank causes the water level to rise 8 in in 1 min. What is the gallons per minute flow to the tank?

51. An 8 in diameter pipe flowing full delivers 510 gpm. What is the feet per second velocity of flow in the pipeline?

52. Express a flow of 10 cfs in terms of gallons per minute.

53. The totalizer reading for the month of December was 134.6 MG. What was the average daily flow (ADF) for the month of September?

54. What is 5.2 MGD expressed as cubic feet per second?

55. A pump discharges into a 3 ft diameter container. If the water level in the container rises 20 in in 30 sec, what is the gallons per minute flow into the container?

57. Convert a flow of 1,825,000 gpd to cubic feet per minute.

58. A 6 in diameter pipeline has water flowing at a velocity of 2.9 fps. What is the gallons per minute flow rate through the pipeline?

59. The velocity in a 10 in pipeline is 2.6 ft/sec. If the 10 in pipeline flows into an 8 in diameter pipeline, what is the feet per second velocity in the 8 in pipeline?

59. Convert 2,225 gpm to gallons per day.

60. The total flow for one day at a plant was 5,350,000 gal. What was the average gallons per minute flow for that day?

9.15   CHEMICAL DOSAGE CALCULATIONS

61. Determine the chlorinator setting (in pounds per day) needed to treat a flow of 5.5 MGD with a chlorine dose of 2.5 mg/L.

62. To dechlorinate a wastewater, sulfur dioxide is to be applied at a level 4 mg/L more than the chlorine residual. What should the sulfonator feed rate be (in pounds per day) for a flow of 4.2 MGD with a chlorine residual of 3.1 mg/L?

63. What should the chlorinator setting be (in pounds per day) to treat a flow of 4.8 MGD if the chlorine demand is 8.8 mg/L and a chlorine residual of 3 mg/L is desired?

64. A total chlorine dosage of 10 mg/L is required to treat the water in a unit process. If the flow is 1.8 MGD and the hypochlorite has 65% available chlorine, how many pounds per day of hypochlorite will be required?

65. The chlorine dosage at a plant is 5.2 mg/L. If the flow rate is 6,250,000 gpd, what is the chlorine feed rate in pounds per day?

66. A storage tank is to be disinfected with 60 mg/L of chlorine. If the tank holds 86,000 gal, how many pounds of chlorine (gas) will be needed?

67. To neutralize a sour digester, 1 lb of lime is to be added for every pound of volatile acids in the digester liquor. If the digester contains 225,000 gal of sludge with a volatile acid (VA) level of 2,220 mg/L, how many pounds of lime should be added?

68. A flow of 0.83 MGD requires a chlorine dosage of 8 mg/L. If the hypochlorite has 65% available chlorine, how many pounds per day of hypochlorite will be required?

9.16 ADVANCED PRACTICE PROBLEMS

1. The diameter of a tank is 60 ft. If the water depth is 20 ft, what is the volume of water in the tank in gallons?

2. A tank is 50 ft in length, 15 ft wide, and 10 ft deep. Calculate the cubic feet volume of the tank.

3. A tank 10 ft wide and 50 ft long is filled with water to a depth of 10 ft. What is the volume of the water in the tank (in gallons)?

4. What is the volume of water in a tank, in gallons, if the tank is 10 ft wide, 40 ft long, and contains water to a depth of 10 ft?

5. A tank has a diameter of 50 ft and a depth of 12 ft. Calculate the volume of water in the tank in gallons.

6. What is the volume of water in a tank, in gallons, if the tank is 20 ft wide, 50 ft long, and contains water to a depth of 16 ft?

7. A rectangular channel is 300 ft in length, 4 ft in depth, and 6 ft wide. What is the volume of water in cubic feet?

8. A replacement section of 10 in pipe is to be sandblasted before it is put into service. If the length of pipeline is 1,500 ft, how many gallons of water will be needed to fill the pipeline?

9. A trapezoidal channel is 700 ft in length, 10 ft wide at the top, 5 ft wide at the bottom, with a distance of 4 ft from top edge to bottom along the sides. Calculate the gallon volume.

10. A section of 8 in diameter pipeline is to be filled with treated water for distribution. If the pipeline is 2,100 ft in length, how many gallons of water will be distributed?

11. A channel is 1,100 ft in length, carries water 4 ft in depth, and 5 ft wide. What is the volume of water in gallons?

12. A pipe trench is to be excavated that is 4 ft wide, 4 ft deep, and 1,100 ft long. What is the volume of the trench in cubic yards?

13. A trench is to be excavated that is 3 ft wide, 4 ft deep, and 400 yd long. What is the cubic yard volume of the trench?

14. A trench is 270 yd long, 3 ft wide, and 3 ft deep. What is the cubic feet volume of the trench?

15. A rectangular trench is 600 ft long, 6.5 ft wide, and 3.5 ft deep. What is the cubic feet volume of the trench?

16. The diameter of a tank is 90 ft. If the water depth in the tank is 20 ft, what is the volume of water in the tank in gallons?

17. A tank is 80 ft long, 20 ft wide, and 12 ft deep. What is the cubic feet volume of the tank?

18. How many gallons of water will it take to fill an 8 in diameter pipe that is 3,000 ft in length?

19. A trench is 500 yd long, 3 ft wide, and 3 ft deep. What is the cubic feet volume of the trench?

20. A trench is to be excavated. If the trench is 3 ft wide, 4 ft deep, and 1,100 ft long, what is the cubic yard volume of the trench?

21. A tank is 30 ft wide and 60 ft long. If the tank contains water to a depth of 12 ft, how many gallons of water are in the tank?

22. What is the volume of water (in gallons) contained in a 2,000 ft section of channel if the channel is 8 ft wide and the water depth is 3.5 ft?

23. A tank has a diameter of 60 ft and a depth of 19 ft. What is the volume of water in the tank in gallons?

24. If a tank is 20 ft in diameter and 30 ft deep, how many gallons of water will it hold?

25. A channel 44 in wide has water flowing to a depth of 2.4 ft. If the velocity of the water is 2 fps, what is the cubic feet per minute flow in the channel?

26. A tank is 20 ft long and 12 ft wide. With the discharge valve closed, the influent to the tank causes the water level to rise 0.7 ft in 1 min. What is the gallons per minute flow to the tank?

27. A trapezoidal channel is 4 ft wide at the bottom and 6 ft wide at the water surface. The water depth is 40 in. If the flow velocity through the channel is 120 ft/min, what is the cubic feet per minute flow rate through the channel?

28. An 8 in diameter pipeline has water flowing at a velocity of 2.2 fps. What is the gallons per minute flow rate through the pipeline? Assume the pipe is flowing full.

29. A pump discharges into a 2 ft diameter container. If the water level in the container rises 28 in in 30 sec, what is the gallons per minute flow into the container?

30. A 10 in diameter pipeline has water flowing at a velocity of 3.0 fps. What is the gallons per minute flow rate through the pipeline if the water is flowing at a depth of 5 in?

31. A channel has a rectangular cross section. The channel is 6 ft wide, with water flowing to a depth of 2.5 ft. If the flow rate through the channel is 14,200 gpm, what is the velocity of the water in the channel (in foot per second)?

32. An 8 in diameter pipe flowing full delivers 590 gpm. What is the velocity of flow in the pipeline (in foot per second)?

33. The velocity in a 10 in diameter pipeline is 2.4 ft/sec. If the 10 in pipeline flows into an 18 in diameter pipeline, what is the velocity in the 8 in pipeline in feet per second?

34. A float travels 400 ft in a channel in 1 min, 32 sec. What is the estimated velocity in the channel (in foot per second)?

35. The velocity in an 8 in diameter pipe is 3.2 ft/sec. If the flow then travels through a 10 in diameter section of pipeline, what is the feet per second velocity in the 10 in pipeline?

36. The following flows were recorded for the week:

    Monday: 4.8 MGD
    Tuesday: 5.0 MGD
    Wednesday: 5.2 MGD
    Thursday: 5.4 MGD
    Friday: 4.8 MGD
    Saturday: 5.2 MGD
    Sunday: 4.8 MGD

    What was the average daily flow rate for the week?

37. The totalizer reading the month of September was 124.4 MG. What was the average daily flow (ADF) for the month of September?

38. Convert 0.175 MGD to gallons per minute.

39. The total flow for one day at a plant was 3,330,000 gal. What was the average gallons per minute flow for that day?

40. Express a flow of 7 cfs in terms of gallons per minute.

41. What is 30 gps expressed as gallons per day?

42. Convert a flow of 4,500,000 gpd to cubic feet per minute.

43. What is 6.5 MGD expressed as cubic feet per second?

44. Express 445,875 cfd as gallons per minute.

45. Convert 2,450 gpm to gallons per day.

46. A channel has a rectangular cross section. The channel is 6 ft wide, with water flowing to a depth of 2 ft. If the flow rate through the channel is 14,800 gpm, what is the velocity of the water in the channel (in foot per second)?

47. A channel 55 in wide has water flowing to a depth of 3.4 ft. If the velocity of the water is 3.5 fps, what is the cubic feet per minute flow in the channel?

48. The following flows were recorded for the months of June, July, and August: June, 107.4 MG; July, 126.8 MG; August, 144.4 MG. What was the average daily flow for this three-month period?

49. A tank is 10 ft by 10 ft. With the discharge valve closed, the influent to the tank causes the water level to rise 8 in in 1 min. What is the gallons per minute flow to the tank?

50. An 8 in diameter pipe flowing full delivers 510 gpm. What is the feet per second velocity of flow in the pipeline?

51. Express a flow of 11 cfs in terms of gallons per minute.

52. The totalizer reading for the month of December was 134.6 MG. What was the average daily flow (ADF) for the month of September?

53. What is 5 MGD expressed as cubic feet per second?

54. A pump discharges into a 3 ft diameter container. If the water level in the container rises 20 in in 30 sec, what is the gallons per minute flow into the container?

55. Convert a flow of 1,820,000 gpd to cubic feet per minute.

56. A 6 in diameter pipeline has water flowing at a velocity of 2.9 fps. What is the gallons per minute flow rate through the pipeline?

57. The velocity in a 10 in pipeline is 2.6 ft/sec. If the 10 in pipeline flows into an 8 in diameter pipeline, what is the feet per second velocity in the 8 in pipeline?

58. Convert 2,220 gpm to gallons per day.

59. The total flow for one day at a plant was 5,300,000 gal. What was the average gallons per minute flow for that day?

Appendix A: Basic Math Answers

1. 660.65
2. 9.84268
3. 0.91
4. 8.5
5. 37.7
6. 0.75
7. 0.167
8. 0.375
9. 13/100
10. 9/10
11. 3/4
12. 49/200
13. 15%
14. 122%
15. 166%
16. 57.1%
17. 0
18. $x - 6 = 2$
    $x - 6 + 6 = 2 + 6$
    $x = 8$
19. $x - 4 = 9$
    $x = 13$
20. $x\ 17 + 8$
    $= 25$
21. $x = 15 - 10$
    $x = 5$
22. $3x/3 = (3)(2)$
    $x = 6$
23. $(4)x/4 = (4)(10)$
    $x = 40$
24. $x = 8/4$
    $x = 2$
25. $x = 15/6$
    $x = 2.5$
26. $x + 10 - 10 = 2 - 10$
    $x = -8$
27. $x - 2 + 2 = -5 + 2$
    $x = -3$

28. $x = -8 - 4$
    $x = -12$
29. $x = -14 + 10$
    $x = -4$
30. $0.5x = -5$
    $x = -10$
31. $9x = -1$
    $x = -1/9$
32. $x^2 = 6^2 = 6 \times 6 = 36$
33. $3 \times 3 \times 3 \times 3 = 81$
34. 1
35. 36:1
36. $2(15) = 5x$
    $30 = 5(x)$
    $6 = x$
37. 3 ft × 3 ft = 9 sq ft
38. 5 in × 3 in = 15 sq in
39. 5 yd × 1 yd = 5 sq yd
40. = 22/7 × 14/1 ft
    = 22/1 × 2/1 ft
    = 44 ft
41. 8 in = 22/7 D
    56 in = 22D
    2.5 in = D
42. Area = πr^2
    = 22/7 × 6^2 in
    = 22/7 × 36 sq in/1
    = 792/7
    = 113 − 1/7 sq in
43. Area = πr^2
    = 22/7 × 5^2 in/1
    = 22/7 × 25 sq/in/1
    = 550 sq in/7
    = 78 4/7 sq inches

Water Treatment Calculations Answer Key

1. 98 ft − 91 ft = 7 ft drawdown
2. 125 ft − 110 ft = 15 ft drawdown
3. 161 ft − 144 ft = 17 ft drawdown
4. (3.7 psi) (2.31 ft/psi) = 8.5 ft water depth in sounding line
   = 112 ft − 8.5 ft
   = 103.5 ft
   = 103.5 ft − 86 ft
   = 17.5 ft

5. (4.6 psi) (2.31 ft/psi) = 10.6 water depth in sounding line
   = 150 ft − 10.6 ft
   = 139.4 ft
   = 171 ft − 139.4 ft
   = 31.4 ft drawdown
6. 300/20 = 15 gpm/ft of drawdown
7. 420 gal/5 min = 84 gpm
8. 810 gal/5 min = 162 gpm
9. 856 gal/5 min = 171 gpm
   (171 gpm) (60 min/hr) = 10,260 gph
10. $\dfrac{(0.785)\ (1\text{ ft})\ (1\text{ ft})\ (12\text{ ft})\ (7.48\text{ gal/cu ft})\ (12\text{ round trips})}{5\text{ minutes}} = 169\text{ gpm}$
11. 750 gal/5 min = 150 gpm
    (150 gpm) (60 min/hr) = 9,000 gph
    (9,000 gph) (10 hrs/day) = 90,000 gal/day
12. 200 gpm/28 ft = 7.1 gpm/ft
13. 620 gpm/21 ft = 29.5 gpm/ft
14. 1,100 gpm/41.3 ft = 26.6 gpm/ft
15. $\dfrac{x\text{ gpm}}{42.8\text{ ft}} = 33.4$ fpm/ft $x = (33.4)\ (42.8)$
    $x = 1,430$ gpm
16. (0.785) (0.5 ft) (0.5 ft) (140 ft) (7.48 gal/cu ft) = 206 gal
    (40 mg/L) (0.000206 MG) (8.34 lb/gal) = 0.07 lb chlorine
17. (0.785) (1 ft) (1 ft) (109 ft) (7.48 gal/cu ft) = 640 gal
    (40 mg/L) (0.000640 MG) (8.34 lb/gal) = 0.21 lb chlorine
18. (0.785) (1 ft) (1 ft) (109 ft) (7.48 gal/cu ft) = 633 gal
    (0.785) (0.67 ft) (0.67 ft) (40 ft) (7.48 gal/cu ft) = 105 gal
    633 + 105 gal = 738 gal
    (110 mg/L) (0.000738 gal) (8.34 lb/gal) = 0.68 lb chlorine
19. (x mg/L) (0.000540 gal) (8.34 lb/gal) = 0.48 lb
    $x = \dfrac{0.48}{(0.000540)\ (8.34)}$
    $x = 107$ mg/L
20. $\dfrac{0.09\text{ lb chlorine}}{5.25/100} = 1.5$ lb $\dfrac{1.5\text{ lb}}{8.34\text{ lb/gal}} = 0.18$ gal
    (0.18 gal) (128 fluid oz/gal) = 23 fl oz
21. (0.785) (0.5 ft) (0.5 ft) (120 ft) (7.48 gal/cu ft) = 176 gal
    $\dfrac{(50\text{ mg/L chlorine})\ (0.000176\text{ MG})\ (8.34\text{ lb/gal})}{65/100} = 0.1$ calcium hypochlorite
    (0.1 lb) (16 oz/1lb) = 1.6 oz calcium hypochlorite
22. (0.785) (1.5 ft) (1.5 ft) (105 ft) (748 gal/cu ft) = 1,387 gal
    $\dfrac{(100\text{ mg/L})\ (0.001387\text{ MG})\ (8.34\text{ lb/gal})}{25/100} = 4.6$ lb chloride of lime

23. $\dfrac{(60 \text{ mg/L}) \ (0.000240 \text{ MG}) \ (8.34 \text{ lb/gal})}{5.25/100} = 2.3 \text{ lb} \quad \dfrac{2.3 \text{ lb}}{8.34 \text{ lb/gal}} = 0.3 \text{ gal}$

    $(0.3 \text{ gal}) \ (128 \text{ fl oz/gal}) = 38.4 \text{ fl oz sodium hypochlorite}$

24. $(4.0 \text{ psi}) \ (2.31 \text{ ft/psi}) = 9.2 \text{ ft}$

25. $(94 \text{ ft} + 24 \text{ ft}) + (3.6 \text{ psi}) \ (2.31 \text{ ft/psi})$
    $= 118 \text{ ft} + 8.3 \text{ ft}$
    $= 126.3 \text{ ft}$

26. $(4.2 \text{ psi}) \ (2.31 \text{ ft/psi}) = 9.7 \text{ ft}$
    $180 \text{ ft} + 9.7 \text{ ft} = 189.7 \text{ ft}$

$$whp = \dfrac{(189.7 \text{ ft}) \ (800 \text{ gpm})}{3960}$$

    $= 38.3 \text{ whp}$

27. $(4.4 \text{ psi}) \ (2.31 \text{ ft/psi}) = 10.2 \text{ ft}$
    Field head $= 200 \text{ ft} + 10.2 \text{ ft} = 210.2 \text{ ft}$

$$Whp = \dfrac{(210.2 \text{ ft}) \ (1000 \text{ gpm})}{3960}$$

    $= 53 \text{ whp}$

28. $\dfrac{(184 \text{ ft}) \ (700 \text{ gpm})}{(3960) \ 83/100} = 39 \text{ bowl bhp}$

29. Shaft friction loss $= 0.67$

$$\dfrac{(0.67 \text{ hp loss})}{100 \text{ ft}} (181 \text{ ft}) = 1.2 \text{ hp loss}$$

    Field hp $= 59.5 \text{ bhp} + 1.2 \text{ hp}$
            $= 60.7 \text{ bhp}$

30. $mhp = \dfrac{\text{Total bhp}}{\text{Motor Effic}/100} \quad \dfrac{58.3 \text{ bhp} + 0.5 \text{ hp}}{90/100} = 65.3 \text{ mhp}$

31. $\dfrac{45 \text{ hp}}{56.4 \text{ bhp}} \times 100 = 80\%$

32. $\dfrac{55.7 \text{ bhp}}{90/100} = 62 \text{ hp input} \quad \dfrac{43.5 \text{ whp}}{62 \text{ Input hp}} = 70\%$

33. $(400 \text{ ft}) \ (110 \text{ ft}) \ (14 \text{ ft}) \ (7.48 \text{ gal/cu ft}) = 4{,}607{,}680 \text{ gal}$

34. $(400 \text{ ft}) \ (110 \text{ ft}) \ (30 \text{ ft} \times 0.4 \text{ average depth}) \ (7.48 \text{ gal/cu ft}) = 3{,}949{,}440 \text{ gal}$

35. $\dfrac{(200 \text{ ft}) \ (80 \text{ ft}) \ (12 \text{ ft})}{43{,}560 \text{ cu ft/ac-ft}} = 4.4 \text{ ac-ft}$

36. $\dfrac{(320 \text{ ft}) \ (170 \text{ ft}) \ (16 \text{ ft}) \ (0.4)}{43{,}560 \text{ cu ft/ac-ft}} = 8.0 \text{ ac-ft}$

37. $\dfrac{(0.5 \text{ mg/L chlorine}) \ (20 \text{ MG}) \ (8.34 \text{ lb/gal})}{25/100} = 334 \text{ lb copper sulfate}$

38. $(62 \text{ ac-ft}) \ (43{,}560 \text{ cu ft/ac-ft}) \ (7.48 \text{ gal/cu ft}) = 20{,}201{,}385 \text{ gal}$

$$\frac{(0.5 \text{ mg/L}) \ (20.2 \text{ MG}) \ (8.34 \text{ lb/gal})}{0.25} \ 337 \text{ lb copper sulfate}$$

39. $\dfrac{1.1 \text{ lb CuSO}_4}{1 \text{ ac-ft}} (38 \text{ ac-ft}) = 41.8 \text{ lb copper sulfate}$

40. Volume, ac-ft $= \dfrac{(250 \text{ ft}) \ (75 \text{ ft}) \ (10 \text{ ft})}{43,560 \text{ cu ft/ac-ft}} = 4.3 \text{ ac-ft}$

$$\frac{(0.8 \text{ lb CuSO}_4)}{1 \text{ ac-ft}} (4.3 \text{ ac-ft}) = 3.14 \text{ lb copper sulfate}$$

41. Volume, ac-ft $= \dfrac{(500 \text{ ft}) \ (100 \text{ ft})}{43,560 \text{ sq ft/ac}} = 1.1 \text{ ac}$

$$\frac{(5.1 \text{ lb CuSO}_4)}{1} (1.1 \text{ ac}) = 5.9 \text{ lb copper sulfate}$$

42. 131.9 ft – 93.5 ft = 38.4 ft

43. $\dfrac{707 \text{ gallons}}{5 \text{ minutes}} = 141 \text{ gpm } (141 \text{ gpm}) (60 \text{ min/hr}) = 8,460 \text{ gph}$

44. $\dfrac{(0.785) \ (1 \text{ ft}) \ (1 \text{ ft}) \ (12 \text{ ft}) \ (7.48 \text{ gal/cu ft}) \ (8 \text{ round trips})}{5 \text{ gpm}} = 113 \text{ gpm}$

45. (3.5 psi) (2.31 ft/psi) = 8.1 water depth in sounding line
     = 167 ft – 8.1 ft
     = 158.9 ft pumping water level, ft
     Drawdown, ft = 158.9 ft – 141 ft
     = 17.9 ft drawdown

46. $\dfrac{610 \text{ gpm}}{28 \text{ ft drawdown}} = 21.8 \text{ gpm/ft}$

47. (0.785) (0.5 ft) (0.5 ft) (150 ft) (7.48 gal/cu ft) = 220 gal
     Pounds chlorine required: (55 mg/L) (0.000220 MG) (8.34 lb/gal) = 0.10 lb chlorine

48. 780 gal/5 min = 156 gpm
     (156 gal/min) (60 min/hr) (8 hrs/day) = 74,880 gal/day

49. (x mg/L) (0.000610 MG) (8.34 lb/gal) = 0.47 lb
     $x = \dfrac{0.47}{(0.000610) \ (8.34)}$
     = 92.3 mg/L

50. (0.785) (1 ft) (1 ft) (89) (7.48 gal/cu ft) = 523 gal
     (0.785) (0.67) (0.67) (45 ft) (7.48 gal/cu ft) = 119 gal
     523 gal + 119 gal = 642 gal
     (100 mg/L) (0.000642 MG) (8.34 lb/gal) = 0.54 lb chlorine

51. $\dfrac{0.3 \text{ lb chlorine}}{5.25/100} = 5.7 \text{ lb} \ \dfrac{5.7 \text{ lb}}{8.34 \text{ lb/gal}} = 0.68 \text{ gal}$
     (0.68 gal) (128 fl oz/gal) = 87 fl oz

52. (4.5 psi) (2.31 psi) = 10.4 ft
53. (3.6 psi) (2.31 ft/psi) = 8.3 ft
    95 ft + 25 ft + 8.3 ft = 128.3 ft field head
54. Field HD = 191 ft + (4.1 psi) (2.31 ft/psi)
    191 ft + 9.5 ft = 200.5 ft

$$\text{whp} \frac{(200.5 \text{ ft}) \ (850 \text{ gpm})}{3960} = 43 \text{ whp}$$

55.

$$= \frac{(175 \text{ ft}) \ (800 \text{ gpm})}{(3960) \ (0.80)} = 44.2 \text{ whp}$$

56. $= \dfrac{47.8 \text{ bhp} + 0.8 \text{ hp}}{0.90} = 54 \text{ hp input}$

57. $\dfrac{45.6 \text{ hp}}{57.4 \text{ bhp}} \times 100 = 79.8\%$

58. $\dfrac{54.7 \text{ bhp}}{0.90} = 61 \text{ hp input}$ Overall Efficiency, $\% = \dfrac{44.6 \text{ whp}}{61 \text{ input hp}} = 73\%$

59. (53 ac-ft) (43,560 cu ft/ac ft) = 2,308,680 cu ft
    = (2,308,680 cu ft) (7.48 gal/cu ft) = 17,268,926 gal

$$= \frac{(0.5 \text{ mg/L}) \ (17.2 \text{ MG}) \ (8.34 \text{ lb/gal})}{0.25}$$

    = 287 lb copper sulfate

60. Area, acres $= \dfrac{(440 \text{ ft}) \ (140 \text{ ft})}{43,560 \text{ sq ft/ac}} = 1.4 \text{ ac}$ (5.5 lb copper sulfate) (1.4 ac) =

    7.7 lb copper sulfate/ac
61. Volume, gal = (4 ft) (5 ft) (3 ft) (7.48 gal/cu ft) = 449 gal
62. Volume, gal = (50 ft) (20 ft) (8 ft) (7.48 gal/cu ft) = 59,840 gal
63. Volume, gal = (40 ft) (16 ft) (8 ft) (7.48 gal/cu ft) = 38,298 gal
64. 42 in/12 in/ft = 3.5 ft
    Volume, gal = (5 ft) (5 ft) 3.5 ft) (7.48 gal/cu t) = 655 gal
65. 2 in/12 in/ft = 0.17 ft
    Volume, gal = (40 ft) 25 ft) (9.17 ft) (7.48 gal/cu ft) = 68,592 gal
66. $\dfrac{3,625,000 \text{ gpd}}{1440 \text{ min/day}} = 2517 \text{ gpm}$

$$\frac{(60 \text{ ft}) \ (25 \text{ ft}) \ (9 \text{ ft}) \ (7.48 \text{ gal/cu ft})}{2517 \text{ gpm}} = 40.1 \text{ min}$$

67. $\dfrac{2,800,000 \text{ gpd}}{1440 \text{ min/day}} = 1944 \text{ gpm}$

    Detention time, min. $= \dfrac{(50 \text{ ft}) \ (20 \text{ ft}) \ (8 \text{ ft}) \ (7.48 \text{ gal/cu ft})}{1944 \text{ gpm}} = 30.8 \text{ min}$

68. $\dfrac{9{,}000{,}000 \text{ gpd}}{(1440 \text{ min/day})\ (60 \text{ sec/min})} = 104.2 \text{ gps}$

Detention time, sec $= \dfrac{(6 \text{ ft})\ (5 \text{ ft})\ (5 \text{ ft})\ (7.48 \text{ gal/cu ft})}{104.2 \text{ gps}} = 10.8 \text{ sec}$

69. $\dfrac{2{,}250{,}000 \text{ gpd}}{1440 \text{ min/day}} = 1563 \text{ gpm}$

Detention Time, sec $= \dfrac{(50 \text{ ft})\ (20 \text{ ft})\ 10 \text{ ft})\ (7.48 \text{ gal/cu ft})}{1563 \text{ gpm}} = 47.9 \text{ min}$

70. $\dfrac{3{,}250{,}000 \text{ gpd}}{(1440 \text{ min/day})\ (60 \text{ sec/min})} = 37.6 \text{ gps}$

Detention Time, sec. $= \dfrac{(4 \text{ ft})\ 4 \text{ ft})\ (3.5 \text{ ft})\ (7.48 \text{ gal/cu ft})}{37.6 \text{ gps}} = 11.1 \text{ sec}$

71. (10 mg/L) (3.45 MGD) (8.34 lb/gal) = 288 lb/day

72. (12 mg/L) (1.660 MGD) (8.34 lb/gal) = 166 lb/day

73. (10 mg/L) (2.66 MGD) (8.34 lb/gal) = 222 lb/day

74. (9 mg/L) (0.94 MGD) (8.34 lb/gal) = 71 lb/day

75. (12 mg/L) (4.10 MGD) (8.34 lb/gal) = 410 lb/day

76. (7 mg/L) (1.66 MGD) (8.34 lb/gal) = 97 lb/day dry alum

$\dfrac{97 \text{ lb/day dry alum}}{5.24 \text{ alum/gal solution}} = 18.5 \text{ gpd Alum solution}$

77. (12 mg/L) (3.43 MGD) (8.34 lb/gal) = 550,000 mg/L (x MGD) (8.34 lb/gal)

$\dfrac{(12 \text{ mg/L})\ (3.43)\ (8.34)}{(550{,}000)\ (8.34)} = \text{x MGD}$

x = 0.0000748 MGD

gpd = 74.8 gpd alum solution

78. (10 mg/L) (4.13 MGD) (8.34 lb/gal) = 344 lb/day dry alum

$\dfrac{344 \text{ lb/day Dry Alum}}{5.40 \text{ lb Alum/gal Solution}} = 64 \text{ gpd Solution}$

79. (11 mg/L) (0.88 MGD) (8.34 lb/gal) = (550,000 mg/L) (x MGD) (8.34 lb/gal)

$\dfrac{(11)\ (0.88)\ (8.34)}{(550{,}000)\ (8.34)} = \text{x MGD}$

0.0000176 MGD = x

= 17.6 gpd alum solution

80. $\dfrac{640 \text{ mg Alum}}{1 \text{ mL Solution}} \times \dfrac{1000}{1000} = \dfrac{640{,}000 \text{ mg Alum}}{1000 \text{ mL}} = 640{,}000 \text{ mg/L}$ (10 mg/L) (1.85

MGD) (8.34 lb/gal) = (640,000 mg/L) (x MGD) (8.34 lb/gal)

$\dfrac{(10)\ (1.85 \text{ MGD})\ (8.34)}{(640{,}000)\ (8.34)} = \text{x MGD}$

0.0000289 MGD = x

28.9 gpd = x

81. $\dfrac{(40\text{ gpd})\ (3785\text{ mL/gal})}{1440\text{ min/day}} = 105\text{ mL/min}$

82. $\dfrac{(34.2\text{ gpd})\ (3785\text{ mL/gal})}{1440\text{ min/day}} = 90\text{ mL/min}$

83. $(10\text{ mg/L})\ (2.88)\ (8.34\text{ lb/gal}) = (550{,}000\text{ mg/L})\ (x\text{ MGD})\ (8.34\text{ lb/gal})$

$\dfrac{(10)\ (2.88\text{ MGD})\ (8.34)}{(550{,}000)\ (8.34)} = x\text{ MGD}$

$0.0000523\text{ MGD} = x$

$52.4\text{ gpd} = x$

$\dfrac{(52.4)\ (3785\text{ mL/gal})}{1440\text{ min/day}} = 138\text{ mL/min}$

84. $(6\text{ mg/L})\ (282\text{ MGD})\ (8.34\text{ lb/gal}) = (550{,}000\text{ mg/L})\ (x\text{ MGD})\ (8.34\text{ lb/gal})$

$\dfrac{(6)\ (2.82)\ (8.34)}{(550{,}000)\ (8.34)} = x\text{ MGD}$

$0.0000307\text{ MGD} = x$

$30.7\text{ gpd} = x$

$\dfrac{(30.7\text{ gpd})\ (3785\text{ mL/gal})}{1440\text{ min/day}} = 80.7\text{ mL/min}$

85. $\dfrac{(10\text{ mg/L})\ (3.45\text{ MGD})\ (8.34\text{ lb/gal})}{5.40\text{ alum/gal solution}} = 53.3\text{ mL/min}$

$\dfrac{(53.3\text{ gpd})\ (3785\text{ mL/gal})}{1440\text{ min/day}} = 141\text{ mL/min}$

86. $140\text{ g} = (0.0022\text{ lb})\ (140)$

$= 0.31\text{ lb dry polymer}$

$\dfrac{0.31\text{ lb}}{(16\text{ gal})\ (8.34\text{ lb/gal}) + 0.31\text{ lb}} \times 100 = 0.23\%$

87. $22\text{ oz/16 oz/lb} = 1.38\text{ lb dry polymer}$

$\dfrac{1.38\text{ lb}}{(24\text{ gal})\ (8.34\text{ lb/gal}) + 1.38\text{ lb}} \times 100 = 0.68\%$

88. $\dfrac{2.1\text{ lb }(100)}{(x\text{ gal})\ (8.34\text{ lb/gal}) + 2.1\text{ lb}} = 0.8 \quad \dfrac{210}{8.34 + 2.1} = 0.8$

$210 = (0.8)\ (8.34x + 2.1)$

$210 = 6.7x + 1.7$

$208.3 = 6.7x$

$31\text{ gal} = x$

89. $0.11x = (0.005)\ (60)$

$x = \dfrac{(0.005)\ (160)}{0.11}$

$x = 7.27\text{ lb liquid polymer}$

90. $(1.3)\ (8.34\ \text{lb/gal}) = 10.8\ \text{lb/gal}$

   8/100 (x gal liquid polymer) (10.8 lb/gal) = 0.2/100 (50 gal lb/gal) (8.34 lb/gal)

   $(0.08)\ (x)\ (10.8) = (0.002)\ (50)\ (8.34)$

   $$x = \frac{(0.002)\ (50)\ (8.34)}{(0.09)\ (10.8)}$$

   $x = 0.86$ gal liquid polymer

91. 11/100 (x gal) (10.1 lb/gal) = 0.8/100 (80 gal) (8.34 lb/gal)

   $(0.11)\ (x)\ (10.1) = (0.008)\ (80)\ (8.34)$

   $$x = \frac{(0.008)\ (80)\ (8.34)}{(0.11)\ (10.1)}$$

   $x = 4.9$ gal liquid polymer

92. $$\frac{10/100\ (32\ \text{lb}) + 0.5/100\ (66\ \text{lb})}{32\ \text{lb} + 66\ \text{lb}} \times 100 = \frac{3.2\ \text{lb} + 0.33\ \text{lb}}{98\ \text{lb}} \times 100$$

   $= 3.6\%$

93. $$\frac{15/100\ (5\ \text{gal})\ (11.2\ \text{lb/gal}) + .20/100\ (40\ \text{gal})\ (8.34\ \text{lb/gal})}{(5\ \text{gal})\ (11.2\ \text{lb/gal}) + (40\ \text{gal})\ (8.34\ \text{lb/gal})} \times 100$$

   $$\frac{8.4\ \text{lb} + 0.67\ \text{lb}}{56\ \text{lb} + 334\ \text{lb}} \times 100$$

   $$\frac{1.1\text{lb}}{390\,\text{lb}} \times 100$$

   $= 2.3\%$ strength

94. $$\frac{12/100\ (12\ \text{gal})\ (10.5\ \text{lb/gal}) + 0.75/100\ (50\ \text{gal})\ ((8.40/\text{gal})}{(12\ \text{gal})\ (10.5\ \text{lb/gal}) + (40\ \text{gal})\ (8.40\ \text{lb/gal})} \times 100$$

   $$= \frac{15.1\ \text{lb} + 3.2\ \text{lb}}{126\ \text{lb} + 336\ \text{lb}}$$

   $$\frac{18.3\ \text{lb}}{458\ \text{lb}} \times 100$$

   $= 4.0\ \%$ strength

95. $$\frac{2.3\ \text{lb}}{30\ \text{min}} = 0.08\ \text{lb/min}\ (0.08\ \text{lb/min})\ (1{,}440\ \text{min/day}) = 115.2\ \text{lb/day}$$

96. $$\frac{42\ \text{oz}}{16\ \text{oz lb}} = 2.61\ \text{lb} \qquad \frac{2.6\ \text{lb}}{45\ \text{min}} = 0.06\ \text{lb/min}$$

   $(0.06\ \text{lb/min})\ (1{,}440\ \text{min/day}) = 86.4\ \text{lb/day}$

97. $$\frac{14\ \text{oz}}{16\ \text{oz/lb}} = 0.88\ \text{lb containers}$$

   $$\begin{array}{l} 2.4\ \text{lb chemical} + \text{container} \\ \underline{-.88\text{lb container}} \\ 1.52\ \text{lb chemical} \end{array}$$

   $$\frac{1.52\ \text{lb chemical}}{30\ \text{minutes}} = 0.051\ \text{lb/min}$$

   $(0.051\ \text{lb/min})\ (1{,}440\ \text{min/day}) = 73\ \text{lb/day}$

2.8 lb container + chemical

98. $\dfrac{-0.6 \text{ lb container}}{2.2 \text{ lb chemical}}$ $\qquad \dfrac{2.2 \text{ lb chemical}}{30 \text{ minutes}} = 0.073 \text{ lb/min}$

(0.073 lb/min) (1,440 min/day) = 105 lb/day

99. (x mg/L) (1.92 MGD) (8.34 lb/gal) = 42 lb polymer

$$x = \dfrac{42}{(1.92)\ (8.34)}$$

= 2.6 mg/L

100. (16,000 mg/L) (0.000070 MGD) (8.34 lb/gal) = 9.3 lb/day

101. $\dfrac{590 \text{ mL}}{5 \text{ min}} = 118 \text{ mL/min}$ $\qquad \dfrac{(118 \text{ mL/min})\ (1 \text{ gal})\ (1440 \text{ min/day})}{3785 \text{ mL}} = 44.9 \text{ gpd}$

(12,000 mg/L) (0.0000449 MGD) (8.34 lb/gal) (1.09 sp gr) = 4.9 lb/day

102. $\dfrac{725 \text{ mL}}{5 \text{ min}} = 145 \text{ mL/min}$ $\qquad \dfrac{(145 \text{ mL/min})\ (1 \text{ gal})\ (1440 \text{ min/day})}{3785 \text{ mL}} = 55 \text{ gpd}$

(12,000 mg/L) (0.000055 MGD) (8.34 lb/day) = 5.5 lb/day

103. $\dfrac{950 \text{ mL}}{5 \text{ min}} = 190 \text{ mL/min}$ $\qquad \dfrac{(190 \text{ mL/min})\ (1 \text{ gal})\ (1440 \text{ min/day})}{3785 \text{ mL}} = 72.3 \text{ gpd}$

(14,000 mg/L) (0.0000723 MGD) (8.34 lb/gal) = 8.4 lb/day

104. $\dfrac{1730 \text{ mL}}{10 \text{ min}} = 173 \text{ mL/min}$

$$\dfrac{(173 \text{ mL/min})\ (1 \text{ gal})\ (1440 \text{ min/day})}{3785 \text{ mL}} = 65.8 \text{ gpd}$$

(19,000 mg/L) (0.0000658 MGD) (8.34 lb/gal) (1.09 sp gr) = 11.4 lb/day

105. 4 in/12 in/ft = 0.3 ft

$$\dfrac{(0.785)\ (4 \text{ ft})\ (4 \text{ ft})\ (0.3 \text{ ft})\ (7.48 \text{ gal/cu ft})}{5 \text{ min}} = 5.6 \text{ gpm}$$

106. 4 in/12 in/ft = 0.3 ft

$$\dfrac{(0.785)\ (4 \text{ ft})\ (4 \text{ ft})\ (0.3 \text{ ft})\ (7.48 \text{ gal/cu ft})}{15 \text{ minutes}} = 1.9 \text{ gpm}$$

107. 3 in/12 in/ft = 0.25 ft

$$\dfrac{(0.785)\ (3 \text{ ft})\ (3\text{ft})\ (0.25 \text{ ft})\ (7.48 \text{ gal/cu ft})}{10 \text{ minutes}} = 1.32 \text{ gpm}$$

(1.32 gpm) (1,440 min/day) = 1,901 gpd

108. 2 in/12 in/ft = 0.17 ft

$$\dfrac{(0.785)\ (3 \text{ ft})\ (3 \text{ ft})\ (0.17 \text{ ft})\ (7.48 \text{ gal/cu ft})}{15 \text{ minutes}} = 0.6$$

(0.6 gpm) (1,440 min/day) = 864 gpd

(0.6 gpm) (1,440 min/day) = 864 gpd

(12,000 mg/L) (0.000864 MGD) (8.34 lb/gal) = 86.5 lb/day

109. $\dfrac{(0.785)\ (4\ \text{ft})\ (4\ \text{ft})\ (0.17\ \text{ft})\ (7.48\ \text{gal/cu ft})}{30\ \text{minutes}} = 0.53\ \text{gpm} (0.145 \quad \text{gpm})$

$(1{,}440\ \text{min/day}) = 209\ \text{gpd}$

$(14{,}500\ \text{mg/L})\ (0.000209\ \text{MGD})\ (8.34\ \text{lb/gal}) = 25\ \text{lb/day}$

110. $535\ \text{lb}/7\ \text{days} = 76.4\ \text{lb/day average}$

111. $\dfrac{2200\ \text{lb}}{90\ \text{lb/day}} = 24.4\ \text{days}$

112. $\dfrac{889\ \text{lb}}{58\ \text{lb/day}} = 15.3\ \text{days}$

113. $(0.785)\ (3\ \text{ft})\ (3\ \text{ft})\ (3.4\ \text{ft})\ (7.48\ \text{gal/cu ft}) = 180\ \text{gal}$

$180\ \text{days}/88\ \text{gpd} = 2\ \text{days}$

114. $(2.8\ \text{mg/L})\ (1.8\ \text{MGD})\ (8.34\ \text{lb/gal}) = 42\ \text{lb/day}$

$(42\ \text{lb/day})\ (30\ \text{days}) = 1{,}260\ \text{lb}$

115. $\dfrac{6{,}100{,}000\ \text{gpd}}{(1440\ \text{min/day})\ (60\ \text{sec/min})} = 71\ \text{gps}$

$= \dfrac{(3\ \text{ft})\ (4\ \text{ft})\ (4\ \text{ft})\ (7.48\ \text{gal/cu ft})}{71\ \text{gps}}$

$= 5\ \text{sec}$

116. Volume, gal $= (50\ \text{ft})\ (20\ \text{ft})\ (9\ \text{ft})\ (7.48\ \text{gal/cu ft})$

$= 67{,}320\ \text{gal}$

117. $(9\ \text{mg/L})\ (4.35\ \text{MGD})\ (8.34\ \text{lb/gal}) = 326\ \text{lb/day}$

118. $(10\ \text{mg/L})\ (3.15\ \text{MGD})\ (8.34\ \text{lb/gal}) = (500{,}000\ \text{mg/L})\ (x\ \text{MGD})\ (8.34\ \text{lb/gal})$

$\dfrac{(10)\ (3.15)\ (8.34)}{(500{,}000)\ (8.34)} = x$

$0.000063\ \text{MGD} = x$

$x = 63.0\ \text{gpd}$

119. $(4\ \text{ft})\ (4\ \text{ft})\ (2\ \text{ft})\ (7.48\ \text{gal/cu ft}) = 239\ \text{gal}$

120. $\dfrac{(45\ \text{gpd})\ (3785\ \text{mL/gal})}{1440\ \text{min/day}} = 118\ \text{mL/min}$

121. $\dfrac{2{,}220{,}000\ \text{gpd}}{1440\ \text{min/day}} = 1542\ \text{gpm}$

$\dfrac{(40\ \text{ft})\ (20\ \text{ft})\ (9.17\ \text{ft})\ (7.48\ \text{gal/cu ft})}{1542\ \text{gpm}} = 36\ \text{min}$

122. $(8\ \text{mg/L})\ (1.84\ \text{MGD})\ (8.34\ \text{lb/gal}) = (600{,}000\ \text{mg/L})\ (x\ \text{MGD})\ (10.2\ \text{lb/gal})$

$\dfrac{(8)\ (1.84)\ (8.34)}{(600{,}000)\ (10.2)} = x\ \text{MGD}$

$0.00002\ \text{MGD}$

$20.0\ \text{gpd} = x$

$\dfrac{(20.0\ \text{gpd})\ (3785\ \text{mL/gal})}{1440\ \text{min/day}} = 52.6\ \text{mL/min}$

123. $\dfrac{(180 \text{ gpd}) \ (3785 \text{ mL/gal})}{1440 \text{ min/day}} = 473.1 \text{ mL/min}$

124. $(6 \text{ mg/L}) \ (0.925 \text{ MGD}) \ (8.34 \text{ lb/gal}) = 46.3 \text{ lb/day}$

125. $\dfrac{2.7 \text{ lb}}{(x \text{ gal}) \ (8.34 \text{ lb/gal}) + 2.7 \text{ lb}} \times 100 = 1.4 \quad \dfrac{270}{8.34 \text{ x} + 2.7} = 1.4$

      $8.34x + 2.7$

      $178.6 = 8.34x + 2.7$

      $175.9 = 8.34x$

      $\dfrac{175.9}{8.34} = x$

      $x = 21.1 \text{ gal}$

126. $\dfrac{(16/100) \ (25 \text{ lb}) + (0.6/100) \ (140 \text{ lb})}{25 \text{ lbs} + 140 \text{ lb}} \times 100 \quad = \dfrac{4 \text{ lb} + 0.84}{165 \text{ lb}} \times 100$

      $= \dfrac{4.84 \text{ lb}}{165 \text{ lb}} \times 100$

      $2.9 \text{ \%}$

127. $\dfrac{4.0 \text{ lb chemical} \times 2}{30 \text{ minutes} \times 2} = \dfrac{8 \text{ lb}}{60 \text{ min}}$ or 8 lb/hr (8 lb/hr) (24 hr/day) = 192 lb/day

128. $\dfrac{\begin{array}{l} 4.2 \text{ lb container} + \text{chemical} \\ -2.0 \text{ lb container} \end{array}}{2.2 \text{ lb chemical}}$ $\dfrac{2.2 \text{ lb} \times 2}{30 \text{ minutes} \times 2} = \dfrac{4.4 \text{ lb}}{60 \text{ min}} = 4.4 \text{ .bus/hr}$

      $(4.4 \text{ lb/hr}) \ (24 \text{ lb/day}) = 105.6 \text{ lb/day}$

129. $190 \text{ g} = (0.0022 \text{ lb}) \ (190)$

      $= 0.42 \text{ lb}$

      $\dfrac{0.42 \text{ lb}}{(25 \text{ gal}) \ (8.34 \text{ lb/gal}) + 0.42 \text{ lb}} \times 100$

      $= \dfrac{0.42}{208.5 + 0.42} \times 100$

      $\dfrac{0.42}{208.9} \times 100$

      $= 0.2 \text{ \%}$

130. $\dfrac{760 \text{ mL}}{5 \text{ min}} = 152 \text{ mL/min} \quad \dfrac{(152 \text{ mL/min}) \ (1440 \text{ min/day})}{3785 \text{ mL/gal}} = 58 \text{ gpd}$

      $(20{,}000 \text{ mg/L} \ (0.000058 \text{ MGD}) \ (8.34 \text{ lb/gal}) = 9.7 \text{ lb/day}$

131. $(14 \text{ mg/L}) \ (4.2 \text{ MGD}) \ (8.34 \text{ lb/gal}) = 490 \text{ lb/day}$

      $\dfrac{490 \text{ lb/day}}{5.66 \text{ lb alum/gal solution}} 86.6 \text{ gpd}$

132. $(0.8/10) \ (210 \text{ lb}) = 17 \text{ lb of 10\% solution}$

      $(9.2/10) \ (210 \text{ lb}) = 193 \text{ lb of water}$

133. 60% solution: (1/60) (175 lb) = 2.9 lb of 60% solution
Water: (59/60) (75 lb) = 172 lb of water

134. 3 in/12 in/ft = 0.25 ft

$$\frac{(0.785)\ (4\ ft)\ (4\ ft)(0.25\ ft)(747\ gal/cu\ ft)}{10\ minutes}=23\ gal/10\ minutes$$

= 2.3 gpm

135. (10/100) (x gal) (10.2 lb/gal) = (0.6/100) (80 gal) (8.34 lb/gal)

$$x=\frac{(0.006)\ (80)\ (8.34)}{(0.1)\ (10.2)}$$

= 3.9 gal

136. 710 mL/5 min = 142 mL/min

$$\frac{(142\ mg/L/min)\ (1440\ min/day)}{3785\ mL/gal}=54\ gpd$$

(9,000 mg/L) (0.0000540 MGD) (8.34 lb/gal) = 4.1 lb/day

137. (6 mg/L) (3.7 MGD) (8.34 lb/gal) = 185 lb/day
(185 lb/day) (30 days) = 5,550 lb

138. $\dfrac{550\ lb}{80\ lb/day}=6.9\ days$

139. (70 ft) (30 ft) (14 ft) (7.48 gal/cu ft)
= 219,912 gal

140. (0.785) (80 ft) (80 ft) (12 ft) (7.48 gal/cu ft)
= 450,954 gal

141. (70 ft) (20 ft) (10 ft) (7.48 gal/cu ft)
= 104,720 gal

142. 50,000 gal = (40 ft) (25 ft) (x ft) (7.48 gal/cu ft)

$$\frac{50,000}{(40)\ (25)\ (7.48)}=x$$

6.7 ft = x

143. 5 in/12 in/ft = 0.42 ft
= (0.785) (75 ft) (75 ft) (10.42 ft) (7.48 gal/cu ft)
= 344,161 gal

144. $\dfrac{2,220,000}{24\ hrs/day}=92,500\ gph$ $\dfrac{(70\ ft)\ (25\ ft)\ (10\ ft)\ (7.48\ gal/cu\ ft)}{92,500\ gph}=1.4\ hrs$

145. $\dfrac{2,920,000\ gpd}{24\ hr/day}=121,667\ gph$

$$\frac{(0.785)\ (80\ ft)\ (80\ ft)\ (12\ ft)\ (7.48\ gal/cu\ ft)}{121,667\ gph}=3.7\ hrs$$

146. $\dfrac{1,520,000\ gpd}{24\ hr/day}=63,333\ gph$

$$\frac{(60\ ft)\ (20\ ft)\ (10\ ft)\ (7.48\ gal/cu\ ft)}{63,333\ gph}=1.4\ hrs$$

147. $3 \text{ hrs} = \dfrac{(0.785)\ (60\text{ ft})\ (60\text{ ft})\ (12\text{ ft})\ (7.48\text{ gal/cu ft})}{x\text{ gph}}$

$x = \dfrac{(0.785)\ (60)\ (60)\ (12)\ (7.48)}{3}$

$x = 84{,}554 \text{ gph}$
$(84{,}554 \text{ gph})\ (24 \text{ hr/day}) = 2{,}029{,}296 \text{ gpd}$
$= 2.0 \text{ MGD}$

148. $\dfrac{1{,}740{,}000 \text{ gpd}}{24 \text{ hr/day}} = 72{,}500 \text{ gph}$

$\dfrac{(70\text{ ft})\ (25\text{ ft})\ (12\text{ ft})\ (7.48\text{ gal/cu ft})}{72{,}500 \text{ gph}} = 2.2 \text{ hr}$

149. $\dfrac{510 \text{ gpm}}{(60\text{ ft})\ (25\text{ ft})} = 0.34 \text{ gpm/sq ft}$

150. $\dfrac{1610 \text{ gpm}}{(0.785)\ (70\text{ ft})\ (70\text{ ft})} = 0.42 \text{ gpm/sq ft}$

151. $\dfrac{540{,}000 \text{ gpd}}{1440 \text{ min/day}} = 375 \text{ gpm}$

$\dfrac{375 \text{ gpm}}{(50\text{ ft})\ (20\text{ ft})} = 0.38 \text{ gpm/sq ft}$

152. $0.5 \text{ gpm/sq ft} = \dfrac{x \text{ gpm}}{(80\text{ ft})\ (25\text{ ft})}$ $(0.5)\ (80)\ (25) = x \text{ gpm}$

$1{,}000 \text{ gpm} = x$
$(1{,}000 \text{ gpm})\ (1{,}440 \text{ min/day}) = 1{,}440{,}000 \text{ gpd}$

153. $\dfrac{1{,}820{,}000 \text{ gpd}}{1440 \text{ min/day}} = 1264 \text{ gpm}$ $\dfrac{1264 \text{ gpm}}{(0.785)\ (60\text{ ft})\ (60\text{ ft})} = 0.45 \text{ gpm/sq ft}$

154. $\dfrac{1{,}550{,}000 \text{ gpd}}{(1440 \text{ min/day})\ (7.48\text{ gal/cu ft})} = 144 \text{ cfm}$ $144 \text{ cfm} = (25\text{ ft})\ (12\text{ ft})\ (x \text{ fpm})$

$\dfrac{144}{(25)(12)} = x$

$x = 0.5 \text{ fpm}$

155. $\dfrac{1{,}800{,}000 \text{ gpd}}{(1440 \text{ min/day})\ (7.48\text{ gal/cu ft})} = 167 \text{ cfm}$ $167 \text{ cfm} = (30\text{ ft})\ (12\text{ ft})\ (x \text{ fpm})$

$\dfrac{167}{(30)(12)} = x$

$x = 0.5 \text{ fpm}$

156. $\dfrac{2{,}450{,}000 \text{ gpd}}{(1440 \text{ min/day})\ (7.48\text{ gal/cu ft}} = 227 \text{ cfm}$ $227 \text{ cfm} = (30\text{ ft})\ (14\text{ ft})\ (x \text{ fpm})$

$\dfrac{227}{(30)(14)} = x$

$x = 0.5 \text{ fpm}$

157. $\dfrac{2,880,000 \text{ gpd}}{(1440 \text{ min/day})\ (7.48 \text{ gal/cu ft})} = 267 \text{ cfm}$ $267 \text{ cfm} = (40 \text{ ft})\ (10 \text{ ft})\ (x \text{ fpm})$

$\dfrac{267}{(40)(12)} = x$

$x = 0.56 \text{ fpm}$

158. $\dfrac{910,000 \text{ gpd}}{(1440 \text{ min/day})\ (7.48 \text{ gal/cu ft})} = 84.5 \text{ cfm}$ $84.5 \text{ cfm} = (25 \text{ ft})\ (10 \text{ ft})\ (x \text{ fpm})$

$\dfrac{84.5}{(25)(10)} = x$

$x = 0.4 \text{ fpm}$

159. $\dfrac{2,520,000 \text{ gpd}}{1440 \text{ min/day}} 1750 \text{ gpm}\ = \dfrac{1750 \text{ gpm}}{(3.14)\ (70 \text{ ft})} = 7.9 \text{ gpm/ft}$

160. $\dfrac{1,890,000 \text{ gpd}}{1440 \text{ min/day}} = 1313 \text{ gpm}$ $\dfrac{1313 \text{ gpm}}{170 \text{ ft}} = 7.7 \text{ gpm/ft}$

161. $\dfrac{1,334,000 \text{ gpd}}{1440 \text{ min/day}} = 926 \text{ gpm}$ $= \dfrac{926 \text{ gpm}}{120 \text{ ft}}$

$= 7.7 \text{ gpm/ft}$

162. $\dfrac{3,700,000 \text{ gpd}}{1440 \text{ min/day}} = 2569 \text{ gpm}$ $\dfrac{2569 \text{ gpm}}{(3.14)\ (70 \text{ ft})} = 11.7 \text{ gpm/ft}$

163. $\dfrac{1,900,000 \text{ gpd}}{1440 \text{ min/day}} = 1319 \text{ gpm}$ $= \dfrac{1319 \text{ gpm}}{160 \text{ ft}}$

$= 8.2 \text{ gpm/ft}$

164. $\dfrac{22 \text{ mL}}{100 \text{ mL}} \times 100$

$= 22\%$

165. $\dfrac{25 \text{ mL}}{100 \text{ mL}} \times 100$

$= 25\%$

166. $\dfrac{15 \text{ mL}}{100 \text{ mL}} \times 100 = 15\%$

167. $\dfrac{16 \text{ mL}}{100 \text{ mL}} \times 100 = 16\%$

168. $\dfrac{0.45 \text{ mg/L Alk}}{1 \text{ mg/L Alum}} = \dfrac{x \text{ mg/L Alk}}{52 \text{ mg/L Alum}}$ $(0.45)\ (52) = x$

$23.4 \text{ mg/L} = x$

$23.4 \text{ mg/L} + 40 \text{ mg/L}$

$= 63.4 \text{ mg/L}$

169. $\dfrac{0.45 \text{ mg/L Alk}}{1 \text{ mg/L Alum}} = \dfrac{x \text{ mg/L Alk}}{60 \text{ mg/L Alum}}$ $(0.45)\ (60) = x$

$x = 27$ mg/L

$\quad = 27$ mg/L $+ 30$ mg/L

$\quad = 57$ mg/L

170. $= 40$ mg/L $- 26$ mg/L

$\quad = 14$ mg/L alk to be added

171. 40 mg/L $- 28$ mg/L

$\quad = 12$ mg/L alk to be added

172. $\dfrac{0.45 \text{ mg/L Alk}}{0.45 \text{ mg/L Lime}} = \dfrac{15 \text{ mg/L Alk}}{x \text{ mg/L Lime}}$ $0.45\,x = (15)\,(0.45)$

$x = \dfrac{(15)\,(0.45)}{0.45}$

$x = 15$ mg/L lime

173. $\dfrac{0.45 \text{ mg/L Alk}}{0.35 \text{ mg/L Lime}} = \dfrac{20 \text{ mg/L Alk}}{x \text{ mg/L Lime}}$ $0.45\,x = (20)\,(0.35)$

$x = \dfrac{(20)\,(0.35)}{0.45}$

$x = 15.6$ mg/L lime

174. $\dfrac{0.45 \text{ mg/L Alk}}{1 \text{ mg/L Alum}} = \dfrac{x \text{ mg/L Alk}}{55 \text{ mg/L Alum}}$ $(0.45)\,(55) = x$

$x = 24.8$ mg/L alk

$\quad = 24.8$ mg/L $+ 30$ mg/L

$\quad = 54.8$ mg/L total alk required

$\quad = 54.8$ mg/L $- 35$ mg/L

$\quad = 19.8$ mg/L alk to be added to the water

$\dfrac{0.45 \text{ mg/L Alk}}{0.35 \text{ mg/L}} = \dfrac{19.8 \text{ mg/L Alk}}{x \text{ mg/L Lime}}$

$0.45\,x = (19.8)\,(0.35)$

$x = \dfrac{(19.8)\,(0.35)}{0.45}$

$x = 15.4$ mg/L lime required

175. $(13.8$ mg/L$)\,(2.7$ MGD$)\,(8.34$ lb/gal$) = 311$ lb/day lime

176. $(12.3$ mg/L$)\,(2.24$ MGD$)\,(8.34$ lb/gal$) = 230$ lb/day lime

177. $(16.1$ mg/L$)\,(0.99$ MGD$)\,(8.34$ lb/gal$) = 133$ lb/day lime

178. $(15$ mg/L$)\,(2.2$ MGD$)\,(8.34$ lb/gal$) = 275$ lb/day lime

179. $\dfrac{(205 \text{ lb/day})\,(453.6 \text{ g/lb})}{1440 \text{ min/day}} = 64.6$ g/min lime

180. $\dfrac{(110 \text{ lb/day})\,(453.6 \text{ g/lb})}{1440 \text{ min/day}} = 34.7$ g/min lime

181. $(12$ mg/L$)\,(0.90$ MGD$)\,(8.34$ lb/gal$) = 90$ lb/day lime

$\dfrac{(90 \text{ lb/day})\,(453.6 \text{ g/lb})}{1440 \text{ min/day}} = 28.4$ g/min lime

182. (14 mg/L) (2.66 MGD) (8.34 lb/gal) = 310.1 lb/day lime

$$\frac{(310.1 \text{ lb/day}) \ (453.6 \text{ g/lb})}{1440 \text{ min/day}} = 97.7 \text{ g/min lime}$$

183. $\dfrac{1{,}550{,}000 \text{ gpd}}{24 \text{ hr/day}} = 64{,}583 \text{ gph}$

$$\frac{(66 \text{ ft}) \ (30 \text{ ft}) \ (12 \text{ ft}) \ (7.48 \text{ gal/cu ft})}{64{,}583} = 2.8 \text{ hrs}$$

184. (70 ft) (30 ft) (7.48 gal/cu ft) = 219,912 gal

185. $\dfrac{1{,}620{,}000 \text{ gpd}}{(1440 \text{ min/day}) \ (7.48 \text{ gal/cu ft})} = 150 \text{ cfm} \quad 150 \text{ cfm} = (25 \text{ ft}) \ (12 \text{ ft}) \ (x \text{ fpm})$

$$\frac{150}{(25) \ (12)} = x$$

$x = 0.5 \text{ fpm}$

186. $\dfrac{635{,}000 \text{ gpd}}{(1440 \text{ min/day})} = 441 \text{ gpm} \quad \dfrac{441 \text{ gpm}}{(40 \text{ ft}) \ (25 \text{ ft})} = 0.44 \text{ gpm/sq ft}$

187. (0.785) (70 ft) (70 ft) (14 ft) (7.48 gal/cu ft) = 402,805 gal

188. $\dfrac{2{,}220{,}000 \text{ gpd}}{1440 \text{ min/day}} = 1542 \text{ gpm} \quad \dfrac{1542 \text{ gpm}}{180 \text{ ft}} = 8.6 \text{ gpm/ft}$

189. $\dfrac{2{,}560{,}000 \text{ gpd}}{24 \text{ hr/day}} = 106{,}667 \text{ gph}$

$$\frac{(0.785) \ (60 \text{ ft}) \ (60 \text{ ft}) \ (10 \text{ ft}) \ (7.48 \text{ gal/cu ft})}{106{,}667 \text{ gph}} = 1.98 \text{ hr}$$

190. $\dfrac{1{,}750{,}000 \text{ gpd}}{(1440 \text{ min/day}) \ (7.48 \text{ gal/cu ft})} = 162 \text{ cfm} \quad 162 \text{ cfm} = (30 \text{ ft}) \ (12 \text{ ft}) \ (x \text{ fpm})$

$$\frac{162}{(30) \ (12)} = x$$

$x = 0.45$

191. $\dfrac{1700 \text{ gpm}}{(0.785) \ (70 \text{ ft}) \ (70 \text{ ft})} = 0.4 \text{ gpm/sq ft}$

192. $\dfrac{3{,}150{,}000 \text{ gpd}}{1440 \text{ min/day}} = 2188 \text{ gpm} \quad = \dfrac{2188 \text{ gpm}}{(3.14) \ (70 \text{ ft})}$

$= 9.9 \text{ gpm}$

193. $2 \text{ hr} = \dfrac{(0.785) \ (60 \text{ ft}) \ (60 \text{ ft}) \ (12 \text{ ft}) \ (7.48 \text{ gal/cu ft})}{x \text{ gph}}$

$$x = \frac{(0.785) \ (60) \ (60) \ (12) \ (7.48)}{2}$$

$x = 126{,}831$

(126,831 gph) (24 hr/day) = 3,043,944 gpd (3.04 MGD)

194. $\dfrac{3{,}250{,}000 \text{ gpd}}{\left(1440 \text{ min/day}\right)\left(7.48 \text{ gal/cu ft}\right)} = 302 \text{ cfm}$ 302 cfm = (30 ft) (14 ft) (x fpm)

$\dfrac{302}{\left(30\right)\left(14\right)} = x$

$x = 0.7$ fpm

195. $\dfrac{26 \text{ mL}}{100 \text{ mL}} \times 100 = 26\%$

196. $\dfrac{1 \text{ mg/L Alum}}{0.45 \text{ mg/L Alk}} = \dfrac{50 \text{ mg/L Alum}}{x \text{ mg/L Alk}}$ (0.45) (50) = x

22.5 mg/L = x

Total alk. required, mg/L = 22.5 mg/L + 30 mg/L
= 52.5 mg/L

197. $0.7 \text{ gpm/sq ft} = \dfrac{x \text{ gpm}}{\left(80 \text{ ft}\right)\left(30 \text{ ft}\right)}$ (0.7) (80) (30) = x

1,680 = x

(1,680 gpm) (1,440 min/day) = 2,419,200 gpd

198. (14.5 mg/L) (2.41 MGD) (8.34 lb/gal) = 291 lb/day

199. $\dfrac{21 \text{ mL}}{100 \text{ mL}} \times 100 = 21\%$ settled sludge

200. $\dfrac{3{,}240{,}000 \text{ gpd}}{1440 \text{ min/day}} = 2250 \text{ gpm}$ $\dfrac{2250 \text{ gpm}}{\left(3.14\right)\left(80 \text{ ft}\right)} = 9.0 \text{ gpm/ft}$

201. 50 mg/L − 30 mg/L = 10 mg/L alk to be added to the water

202. $\dfrac{0.45 \text{ mg/L Alk.}}{1 \text{ mg/L Alum}} = \dfrac{x \text{ mg/L Alk}}{50 \text{ mg/L Alum}}$ (0.45) (50) = x

22.5 mg/L alk = x
= 22.5 mg/L + 30 mg/L
= 52.5 mg/L alk required
= 52.5 mg/L − 33 mg/L
= 19.5 mg/L alk to be added

$\dfrac{0.45 \text{ mg/L Alk.}}{0.35 \text{ mg/L Lime}} = \dfrac{19.5 \text{ mg/L Alk}}{x \text{ mg/L Lime}}$

$x = \dfrac{\left(19.5\right)\left(0.35\right)}{0.45}$

$x = 15.2$ mg/L lime required

203. $\dfrac{\left(192 \text{ lb/day Lime}\right)\left(453.6 \text{ g/lb}\right)}{1440 \text{ min/day}} = 60.5 \text{ g/min}$

204. (16 mg/L) (1.5 MGD) (8.34 lb/gal) = 200 lb/day

205. $\dfrac{\left(\text{Lime, lb/day}\right)\left(453.6 \text{ g/lb}\right)}{1440 \text{ min/day}} = \text{Lime, g/min}$

$$\frac{(14 \text{ mg/L}) \ (2.88 \text{ MGD}) \ (8.34 \text{ lb/gal}) \ (453.6 \text{ g/lb})}{1440 \text{ min/day}}$$

$= 106$ g/min lime

206. $\dfrac{14,200,000 \text{ gal}}{(80 \text{ hrs}) \ (60 \text{ min/lb})} = 2958$ gpm

207. $\dfrac{2,970,000 \text{ gpd}}{1440 \text{ min/day}} = 2063$ gpm

208. $3200 \text{ gpm} = \dfrac{16,000,000 \text{ gal}}{(x \text{ hrs}) \ (60 \text{ min/hr})} \quad x = \dfrac{16,000,000}{(3200) \ (60)}$

    $x = 83$ hrs

209. (45 ft) (22 ft) (1ft/5 min) (7.48 gal/cu ft) = 1,481 gpm

210. 14 in/12 in = 1.17 ft

    (40 ft) (30 ft) (1.17 ft/5 min) (7.48 gal/cu ft) = 2,100 gpm

211. 18 in/12 in = 1.5 ft

    (35 ft) (18 ft) (1.5 ft/6 min.) (7.48 gal/cu ft) =1,178 gpm

212. $\dfrac{1760 \text{ gpm}}{(20 \text{ ft}) \ (18 \text{ ft})} = 4.9$ gpm/sq ft

213. $\dfrac{2,150,000 \text{ gal}}{1440 \text{ min/day}} = 1493$ gpm $\dfrac{1493 \text{ gpm}}{(32 \text{ ft}) \ (18 \text{ ft})} = 2.6$ gpm/sq ft

214. $\dfrac{18,100,000 \text{ gal}}{(71.6 \text{ hr}) \ (60 \text{ min/hr})} = 4213$ gpm $\dfrac{4213 \text{ gpm}}{(38 \text{ ft}) \ (24 \text{ ft})} = 4.6$ gpm/sq ft

215. $\dfrac{14,200,000 \text{ gal}}{(71.4 \text{ hr}) \ (60 \text{ min/hr})} = 3315$ gpm $\dfrac{3315 \text{ gpm}}{(33 \text{ ft}) \ (24 \text{ ft})} = 4.2$ gpm/sq ft

216. $\dfrac{3,550,000 \text{ gpd}}{1440 \text{ min/day}} = 2465$ gpm $\dfrac{2465 \text{ gpm}}{(88 \text{ ft}) \ (22 \text{ ft})} = 2.9$ gpm/sq ft

217. 22 in/12 in/ft = 1.83 ft

    (38 ft) (18 ft) (1.83/5 min) (7.48 gal/cu ft) = 1,873 gpm

    $\dfrac{1873 \text{ gpm}}{(38 \text{ ft}) \ (18 \text{ ft})} 2.7$ gpm/sq ft

218. 21 in/12 in/ft = 1.8 ft

    (33 ft) (24 ft) (1.8 ft/6 min) (7.48 gal/cu ft) = 1,777 gpm

    $\dfrac{1777 \text{ gpm}}{(33 \text{ ft}) \ (24 \text{ ft})} = 2.2$ gpm/sq ft

219. $\dfrac{2,870,000 \text{ gal}}{(20 \text{ ft}) \ (18 \text{ ft})} = 7972$ gal/sq ft

220. $\dfrac{4,180,000 \text{ gallons}}{(32 \text{ ft}) \ (20 \text{ ft})} = 6533$ gal/sq ft

221. $\dfrac{2,980,000 \text{ gal}}{(24 \text{ hr})(18 \text{ ft})} = 6898$ gal/sq ft

222. (3.4 gpm/sq ft) (3,330 min) = 11,322 gal/sq ft

223. (2.6 gpm/sq ft) (60.5 hrs) (60 min/hr) = 9,438 gal/sq ft

224. $\dfrac{3510 \text{ gpm}}{380 \text{ sq ft}} = 9.2$ gpm/sq ft

225. $\dfrac{3580 \text{ gpm}}{(18 \text{ ft})(14 \text{ ft})} = 14.2$ gpm/sq ft

226. $\dfrac{(16 \text{ gpm/sq ft})(12 \text{ in./ft})}{7.48 \text{ gal/cu ft}} = 25.7$ in./min

227. $\dfrac{3650 \text{ gpm}}{(30 \text{ ft}) \ (18 \text{ ft})} = 6.8$ gpm/sq ft

228. $\dfrac{3080 \text{ gpm}}{(18 \text{ ft})(14 \text{ ft})} = 12.2$ gpm/sq ft (12.2 gpm/sq ft) (1.6) = 19.5 in/min rise

229. (6,650 gpm) (6 min) = 39,900 gal

230. (9,100 gpm) (7 min) = 63,700 gal

231. (4,670 gpm) (5 min) = 23,350 gal

232. (6,750 gpm) (6 min) = 40,500 gal

233. 59,200 gal = (0.785) (40 ft) (40 ft) (x ft) (7.48 gal/cu ft)

$\dfrac{59,200}{(0.785)(40)(40)(7.48)} = x$

x = 6.3 ft

234. 62,200 gal = (0.785) (52 ft) (52 ft) (x ft) (7.48 gal/cu ft)

$\dfrac{62,200}{(0.785)(52)(52)(7.48)} = x$

x = 3.9 ft

235. 42,300 gal = (0.785) (42 ft) (42 ft) (x ft) (7.48 gal/cu ft)

$\dfrac{42,300 \text{ gal}}{(0.785)(42)(42)(7.48)} = x$

x = 4.1 ft

236. (7,150 gpm) (7 min) = 50,050 gal

50,050 gal = (0.785) (40 ft) (40 ft) (x ft) (7.48 gal/cu ft)

$\dfrac{50,050}{(0.785)(40)(40)(7.48)} = x$

x = 5.3 ft

237. 8,860 gpm (6 min) = 53,160 gal

53,160 gal = (0.785) (40 ft) (40 ft) (x ft) (7.48 gal/cu ft)

Backwash Pumping Rate, gpm $= \dfrac{53,160}{(0.785)(40)(40)(7.48 \text{ gal/cu ft})} = x$

x = 5.7 ft

238. (19 gpm/sq ft) (42 ft) (22 ft) = 17,556 gpm

239. (20 gpm/sq ft) (36 ft) (26 ft) = 18,720 gpm

240. (16 gpm/sq ft) (22 ft) (22 ft) = 7,744 gpm

241. (24 gpm/sq ft) (26 ft) (22 ft) = 13,728 gpm

242. $\dfrac{74,200 \text{ gal}}{17,100,000 \text{ gal}} \times 100 = 0.43\%$

243. $\dfrac{37,200 \text{ gal}}{6,100,000 \text{ gal}} \times 100 = 0.61\%$ backwash water

244. $\dfrac{59,400 \text{ gal}}{13,100,000 \text{ gal}} \times 100 = 0.45\%$ backwash water

245. $\dfrac{52,350 \text{ gal}}{11,110,000 \text{ gal}} \times 100 = 0.47\%$

246. Mud ball volume = 635 mL – 600 mL = 35 mL

$\dfrac{35 \text{ mL}}{3625 \text{ mL}} \times 100$

= 0.97%

247. Mud ball volume = 535 mL – 510 mL = 25 mL
Total sample volume = (5) (705 mL) = 3,525 mL

$\dfrac{25 \text{ mL}}{3625 \text{ mL}} \times 100$

= 0.7%

248. Mud ball volume = 595 mL – 520 mL = 75 mL
Total sample volume = (5) (705 mL) = 3,525 mL

$\dfrac{75 \text{ mL}}{3625 \text{ mL}} \times 100$

= 2.2%

249. Mud ball volume = 562 mL – 520 mL = 42 mL
Total sample volume = (5) (705 mL) = 3,525 mL

$\dfrac{42 \text{ mL}}{3625 \text{ mL}} \times 100$

= 1.2%

250. $\dfrac{11,400,000 \text{ gal}}{(80 \text{ hr})\ (60 \text{ min/hr})} = 2375$ gpm

251. $\dfrac{3,560,000 \text{ gpd}}{1440 \text{ min/day}} = 2472$ gpm $\quad \dfrac{2472 \text{ gpm}}{(40 \text{ ft})\ (25 \text{ ft})} = 2.5$ gpm

252. $\dfrac{2,880,000 \text{ gal}}{(25 \text{ ft})(25 \text{ ft})} = 4608$ gal/sq ft

253. $2900 \text{ gpm} = \dfrac{14,800,000 \text{ gal}}{(x \text{ hr})(60 \text{ min/hr})} \quad x = \dfrac{14,800,000}{(2900)(60)}$

$x = 85.1$ hr

254. 14 in/12 in = 1.17 ft
= (38 ft) (26 ft) (1.17 ft/5 min) = 231 cfm
(231 cfm) (7.48 gal/cu ft) = 1,728 gpm

255. $\dfrac{3{,}450{,}000 \text{ gal}}{(30 \text{ ft}) \ (25 \text{ ft})} = 4{,}600 \text{ gal/sq ft}$

256. $\dfrac{13{,}500{,}000 \text{ gal}}{(73.8 \text{ hr})(60 \text{ min/hr})} = 3049 \text{ gpm} \quad \dfrac{3049 \text{ gpm}}{(30 \text{ ft}) \ (20 \text{ ft})} = 5.1 \text{ gpm/sq ft}$

257. $\dfrac{3220 \text{ gpm}}{360 \text{ sq ft}} = 8.9 \text{ gpm/sq ft}$

258. (6,350 gpm) (6 min) = 38,100 gal

259. 14 in/12 in = 1.2 ft

   (30 ft) (22 ft) (1.2 ft/5 min) (7.48 gal/cu ft) = 1,185 gpm

   Flow Rate, gpm/sq ft $= \dfrac{1185 \text{ gpm}}{(30 \text{ ft})(22 \text{ ft})} = 1.8 \text{ gpm/sq ft}$

260. 53,200 gal = (0.785) (45 ft) (45 ft) (x ft) (7.48 gal/cu ft)

   $\dfrac{53{,}200}{(0.785)(45)(45)(7.48)} = x$

   $x = 4.5$ ft

261. (3.3 gpm/sq ft) (3,620 min) = 11,946 gal/sq ft

262. 20 in/12 in = 1.7 ft

   (40 ft) (25 ft) (1.7 ft/5 min) (7.48 gal/cu ft) = 2,543 gpm

   Flow Rate, gpm/sq ft $= \dfrac{2543 \text{ gpm}}{(40 \text{ ft}) \ (25 \text{ ft})} = 2.5 \text{ gpm/sq ft}$

263. $\dfrac{3800 \text{ gpm}}{(35 \text{ ft})(25 \text{ ft})} = 4.3 \text{ gpm/sq ft}$

264. (4,500 gpm) (7 min) = 31,500 gal

265. (16 gpm/sq ft) (30 ft) (30 ft) = 14,400 gpm

266. $\dfrac{2800 \text{ gpm}}{(25 \text{ ft})(20 \text{ ft})} = 5.6 \text{ gpm/sq ft} \quad \dfrac{(5.6 \text{ gpm/sq ft})(12 \text{ in./ft})}{7.48 \text{ gal/cu ft}} = 8.9 \text{ in./min}$

267. 18 in/12 in = 1.5 ft

   (30 ft) (25 ft) (1.5/6 min) (7.48 gal/cu ft) = 1,403 gpm

   $\dfrac{1403}{(30 \text{ ft})(25 \text{ ft})} = 1.9 \text{ gpm/sq ft}$

268. (18 gpm/sq ft) (45 ft) (25 ft) = 20,250 gpm

269. $\dfrac{71{,}350 \text{ gal}}{18{,}200{,}000 \text{ gal}} \times 100 = 0.39 \text{ backwash water}$

270. 86,400 gal = (0.785) (35 ft) (35 ft) (x ft) (7.48 gal/cu ft)

   $\dfrac{86{,}400}{(0.785)(35)(35)(7.48)} = 12 \text{ ft}$

271. 527 mL − 500 mL = 27 mL

   $\dfrac{27 \text{ mL}}{3480 \text{ mL}} \times 100$

   $= 0.8 \%$

272. $\dfrac{51{,}200 \text{ gal}}{13{,}800{,}000 \text{ gal}} \times 100 = 0.37\%$

273. $571 - 500 \text{ mL} = 71 \text{ mL}$

$$\dfrac{71 \text{ mL}}{(5)(695 \text{ mL})} \times 100$$

$$= 2\%$$

274. (a) $\dfrac{3{,}700{,}000 \text{ gal}}{36 \text{ hr}/500 \text{ sq ft}} \times \dfrac{1 \text{ hr}}{60 \text{ min}} = 3.4 \text{ gpm/sq ft}$

   (b) $12 \text{ gpm/sq ft} \times 15 \text{ min} \times 500 \text{ sq ft} = 90{,}000 \text{ gal}$

   (c) $\dfrac{90{,}000 \text{ gal}}{3{,}700{,}000 \text{ gal}} \times 100 = 2.4\%$

   (d) $(70 \text{ ft}) - (25 \text{ ft}) = 45 \text{ ft}$

   (e) $\dfrac{3{,}700{,}000}{500 \text{ sq ft}} = 7{,}400 \text{ gal/sq ft}$

275. $(1.8 \text{ mg/L}) (3.5 \text{ MGD}) (8.34 \text{ lb/gal}) = 52.5 \text{ lb/day chlorine}$

276. $(2.5 \text{ mg/L}) (1.34 \text{ MGD}) (8.34 \text{ lb/gal}) = 28 \text{ lb/day}$

277. $(0.785) (1 \text{ ft}) (1 \text{ ft}) (1{,}200 \text{ ft}) (7.48 \text{ gal/cu ft}) = 7{,}046 \text{ gal}$
    $(52 \text{ mg/L}) (0.007046 \text{ MG}) (8.34 \text{ lb/gal}) = 3.1 \text{ lb chlorine}$

278. $(x \text{ mg/L}) (3.35 \text{ MGD}) (8.34 \text{ lb/gal}) = 43 \text{ lb}$

$$x = \dfrac{43}{(3.35)\ (8.34)}$$

$x = 1.5 \text{ mg/L}$

279. 19,222,420 gal
    −18,815,108 gal

407,312 gal/24 hr = 0.407 MGD

$$x = \dfrac{16}{(0.407)(8.34)}$$

$x = 4.7 \text{ mg/L}$

280. $1.6 \text{ mg/L} + 0.5 \text{ mg/L} = 2.1 \text{ mg/L}$

281. $2.9 \text{ mg/L} = x \text{ mg/L} + 0.7 \text{ mg/L}$
    $2.9 - 0.7 = x \text{ mg/L}$
    $x = 2.2 \text{ mg/L}$

282. $2.6 \text{ mg/L} + 0.8 \text{ mg/L} = 3.4 \text{ mg/L}$
    $(3.4 \text{ mg/L}) (3.85 \text{ MGD}) (8.34 \text{ lb/gal}) = 109 \text{ lb/day}$

283. $(x \text{ mg/L}) (1.10 \text{ MGD}) (8.34 \text{ lb/gal}) = 6 \text{ lb/day}$

$$x = \dfrac{6}{(1.10)(8.34)}$$

$x = 0.65 \text{ mg/L}$
$0.8 \text{ mg/L} - 0.65 \text{ mg/L} = 0.15 \text{ mg/L}$
Expected increase in residual was 0.65 mg/L, whereas the actual increase in residual was only 0.15 mg/L. From this analysis it appears the water is not being chlorinated beyond the breakpoint.

284. $(x \text{ mg/L}) (2.10 \text{ MGD}) (8.34 \text{ lb/gal}) = 5 \text{ lb/day}$

$$x = \frac{5}{(210)(8.34)}$$

$x = 0.29 \text{ mg/L}$

$0.5 \text{ mg/L} - 0.4 \text{ mg/L} = 0.1 \text{ mg/L}$

The expected chlorine residual increase (0.29 mg/L) is consistent with the actual increase in chlorine residual (0.1 mg/L); thus, it appears as though the water is not being chlorinated beyond the breakpoint.

285. $\dfrac{48 \text{ lb/day}}{0.65} = 73.8 \text{ lb/day Hypochlorite}$

286. $\dfrac{42 \text{ lb/day}}{0.65} = 64.6 \text{ lb/day Hypochlorite}$

287. $(2.7 \text{ mg/L}) (0.928 \text{ MGD}) (8.34 \text{ lb/gal}) = 21 \text{ lb/day chlorine}$

$\dfrac{21 \text{ lb/day}}{0.65} = 32.3 \text{ lb/day Hypochlorite}$

288. $54 \text{ lb/day} = \dfrac{x \text{ lb/day}}{0.65}$ $(54)(0.65) = x$

$x = 35.1 \text{ lb/day chlorine}$

$(x \text{ mg/L}) (1.512 \text{ MGD}) (8.34 \text{ lb/gal}) = 35.1 \text{ lb/day}$

$$x = \frac{35.1}{(1.512 \text{ MGD})(8.34)} = 2.8 \text{ mg/L}$$

289. $49 \text{ lb/day} = \dfrac{x \text{ lb/day}}{0.65}$ $(49)(0.65) = x$

$31.9 \text{ lb/day chlorine}$

$(x \text{ mg/L}) (3.210 \text{ MGD}) (8.34 \text{ lb/gal}) = 31.9 \text{ lb/day}$

$$x = \frac{31.9}{(3.210)(8.34)}$$

$x = 1.2 \text{ mg/L chlorine}$

290. $\dfrac{36 \text{ lb/day}}{8.34 \text{ lb/gal}} = 4.3 \text{ gpd Hypochlorite}$

291. $(2.2 \text{ mg/L}) (0.245 \text{ MGD}) (8.34 \text{ lb/gal}) = (30,000 \text{ mg/L}) (x \text{ MGD}) (8.34 \text{ lb/gal})$

$$\frac{(2.2)(0.245)(8.34)}{(30,000)(8.34)} = x$$

$0.0000179 \text{ MGD} = x$

$17.9 = x$

292. $\dfrac{2,330,000 \text{ gal}}{7 \text{ days}} = 332,857 \text{ gpd}$

$$\frac{(0.785)(3 \text{ ft})(3 \text{ ft})(2.83 \text{ ft})(7.48 \text{ gal/cu ft})}{7 \text{ days}} = 21.4 \text{ gpd}$$

$(x \text{ mg/L}) (0.332 \text{ MGD}) (8.34 \text{ lb/gal}) = (30,000 \text{ mg/L}) (0.0000214 \text{ MGD})$ (8.34 lb/gal)

$$x = \frac{(30,000)\ (0.0000214)\ (8.34)}{(0.332)\ (8.34)}$$

$x = 1.9$ mg/L

293. (400 gpm) (1,440 min/day) = 576,000 gpd

= 0.576 MGD

(1.8 mg/L) (0.576 MGD) (8.34 lb/gal) = (40,000 mg/L) (x MGD) (8.34 lb/gal)

$$\frac{(1.8)\ (0.576)\ (8.34)}{(40,000)\ (8.34)} = x$$

$x = 0.0000259$ MGD

$x = 25.9$ gpd

294. (2.9 mg/L) (0.955 MGD) (8.34 lb/gal) = (30,000 mg/L) (x MGD) (8.34 lb/gal)

$$\frac{(2.9)\ (0.955)\ (8.34)}{(30,000)\ (8.34)} = x$$

$x = 0.0000923$ MGD

$x = 923$ MGD

295. $\dfrac{(22\ \text{lb})\ (0.65)}{(60\ \text{gal})\ (8.34\ \text{lb/gal})\ +\ (22\ \text{lb})\ (0.65)} \times 100 \quad \dfrac{14.3\ \text{lb}}{500.4\ \text{lb}\ +\ 14.3\ \text{lb}} \times 100$

$\dfrac{14.3\ \text{lb}}{514.7\ \text{lb}} \times 100$

= 2.8% chlorine

296. (320 g) (0.0022 lb/gram) = 0.70 lb hypochlorite

$$\frac{(0.70\ \text{lb})\ (0.65)}{(7\ \text{gal})\ (8.34\ \text{lb/gal})\ +\ (0.70\ \text{lb})\ (0.65)} \times 100$$

$\dfrac{0.46\ \text{lb}}{58.4\ +\ 0.46} \times 100$

$\dfrac{0.46\ \text{lb}}{58.9\ \text{lb}} \times 100$

= 0.79% chlorine

297. $3 = \dfrac{(x\ \text{lb})\ (0.65)}{(65\ \text{gal})\ (8.34\ \text{lb/gal})\ +\ (x\ \text{lb})\ (0.65)} \times 100 \quad 3 = \dfrac{(x)\ (0.65)\ (100)}{542.1\ +\ 0.65\ x}$

$3 = \dfrac{65\ x}{542.1\ +\ 0.65\ x} = 21.7\ x$

$542.8 = 21.7x$

$\dfrac{542.8}{21.7} = x$

$x = 25$ lb

298. (x gal) (8.34) (10/100) = (35 gal) (8.34) (2/100)

$$x = \frac{(35)\ (8.34)\ (0.02)}{(8.34)\ (0.10)}$$

$x = 7$ gal

299. $(x \text{ gal}) (8.34) (13/100) = (110 \text{ gal}) (8.34) (1.2/100)$

$$x = \frac{(110) \ (8.34) \ (0.012)}{(8.34) \ (0.13)}$$

$x = 10.2$ gal

300. $(6 \text{ gal}) (8.34) (12/100) = (x \text{ gal}) (8.34) (2/100)$

$$\frac{(6) \ (8.34) \ (0.12)}{(8.34) \ (0.02)} = x$$

$x = 36$ gal solution

Because 6 gal is liquid hypochlorite, a total of 36 gal − 6 gal = 30 gal water must be added.

301. $\dfrac{(50 \text{ lb}) \ (0.11) \ + \ (220 \text{ lb}) \ (0.01)}{50 \text{ lb} \ + \ 220 \text{ lb}} \times 100 \quad \dfrac{5.5 \text{ lb} \ + \ 2.2 \text{ lb}}{270 \ lb} \times 100$

$\dfrac{7.7 \text{ lb}}{270 \text{ lb}} \times 100$

$= 2.85\%$

302. $\dfrac{(12 \text{ gal}) \ (8.34 \text{ lb/gal}) \ (12/100) \ + \ (60 \text{ gal}) \ (8.34 \text{ lb/gal}) \ (1.5/100)}{(12 \text{ gal}) \ (8.34 \text{ lb/gal}) \ + \ (60 \text{ gal}) \ (8.34 \text{ lb/gal})} \times 100$

$\dfrac{12 \text{ gal} \ + \ 7.5 \text{ lb}}{100 \text{ lb} \ + \ 500 \text{ lb}} \times 100$

$\dfrac{19.5 \text{ lb}}{600 \text{ lb}} \times 100$

$= 3.3\%$

303. $\dfrac{(16 \text{ gal}) \ (8.34 \text{ lb/gal}) \ (12/100) \ + \ (70 \text{ gal}) \ (8.34) \text{ lb/gal}) \ (1/100)}{(16 \text{ gal}) \ (8.34 \text{ lb/gal}) \ + \ (70 \text{ gal}) \ (8.34 \text{ lb/gal})} \times 100$

$\dfrac{16 \text{ lb} \ + \ 5.8 \text{ lb}}{133.4 \text{ lb} \ + \ 583.8 \text{ lb}} \times 100$

$\dfrac{21.8 \text{ lb}}{717.2 \text{ lb}} \times 100$

$= 3.0\%$

304. $\dfrac{1,000 \text{ lb}}{44 \text{ lb/day}} = 22.7$ days

305. 8 in/12 in = 0.67 ft

$$x \text{ days} = \frac{(0.785) \ (4 \text{ ft}) \ 4 \text{ ft}) \ (3.67 \text{ ft}) \ (7.48 \text{ gal/cu ft})}{80 \text{ gpd}}$$

$= 345/80 = 4.3$ days' supply

306. $\dfrac{(24 \text{ lb/day}) \ (150 \text{ hr operation})}{(24 \text{ lb/day})} = 150$ lb chlorine used

307. $\dfrac{(12 \text{ lb/day})}{24 \text{ hr/day}} (111 \text{ hour operation}) = 55.5$ lb

91 lb − 55.5 lb = 35.5 lb remaining

308. (55 lb/day) (30 days) = 1,650 lb chlorine/month

$$\frac{1650 \text{ lb chlorine}}{150 \text{ lb/cylinder}} = 11 \text{ chlorine cylinders}$$

309. $2 \text{ days} = \dfrac{(0.785) \ (3 \text{ ft}) \ 3 \text{ ft} \ (x \text{ ft}) \ (7.48 \text{ gal/cu ft})}{52 \text{ gpd}}$

$$\frac{(20) \ (52)}{(0.785) \ (3) \ (3) \ (7.48)} = x$$

$$x = \frac{104}{52.8}$$

$$x = 2 \text{ ft}$$

310. 1.8 mg/L + 0.9 mg/L = 2.7 mg/L

311. (23 mg/L) (0.98 MGD) (8.34 lb/gal) = 18.8 lb/day

312. $\dfrac{60 \text{ lb/day}}{0.65} = 92.3 \text{ lb/day Hypochlorite}$

313. $\dfrac{51 \text{ lb/day}}{8.34 \text{ lb/gal}} = 6.1 \text{ gpd}$

314. 3.1 mg/L = x mg/L + 0.6 mg/L

    3.1 − 0.6 = x

    x = 2.5 mg/L

315. $\dfrac{(30 \text{ lb}) \ (0.65)}{(66 \text{ gal}) \ (8.34 \text{ lb/gal}) \ + \ (30 \text{ lb}) \ (0.65)} \times 100$

$$\frac{30.65 \text{ lb}}{550.4 \text{ lb} \ + \ 19.5 \text{ lb}} \times 100$$

$$\frac{30.65 \text{ lb}}{569.9 \text{ lb}} \times 100$$

$$= 5.4\%$$

316. (1,620 gpm) (1,440 min/day) = 2,332,800 gpd

    (2.8 mg/L) (2.332 MGD) (8.34 lb/gal) = 54.5 lb/day

317. (2.8 mg/L) (1.33 MGD) (8.34 lb/gal) = (12,500 mg/L) (x MGD) (8.34 lb/gal)

$$\frac{(2.8) \ (1.33) \ (8.34)}{(12,500) \ (8.34)} = x$$

    0.0002979 MGD = x

    297.9 gpd = x

318. Volume, gal = (0.785) (0.67) (0.67) (1,600 ft) (7.48 gal/cu ft) = 4,217 gal

    (60 mg/L) (0.004217 MG) (8.34 lb/gal) = 2.1 lb

319. (x mg/L) (2.11 MGD) (8.34 lb/gal) = 3 lb/day

$$x = \frac{3}{(2.11) \ (8.34)}$$

$$x = 0.17 \text{ mg/L}$$

0.6 mg/L – 0.5 mg/L = 0.1

Chlorination is assumed to be at the breakpoint.

320. $\dfrac{(70 \text{ gal}) \ (8.34 \text{ lb/gal}) \ (12/100) \ + \ (250 \text{ gal}) \ (8.34 \text{ lb/gal}) \ (2/100)}{(70 \text{ gal}) \ (8.34 \text{ lb/gal}) \ + \ (250 \text{ gal}) \ (8.34 \text{ lb/gal})} \times 100$

$\dfrac{70 \text{ lb} + 41.7}{584 + 2085} \times 100$

$\dfrac{111.7 \text{ lb}}{2669 \text{ lb}} \times 100$

$= 4.2\%$

321. $\dfrac{310 \text{ lb}}{34 \text{ lb/day}} = 9.1 \text{ days}$

44,115,670 gal

322. $\dfrac{-43,200,000 \text{ gal}}{915,670 \text{ gal}}$ $(x \text{ mg/L}) \ (0.915 \text{ MGD}) \ (8.34 \text{ lb/gal}) = 18 \text{ lb/day}$

$x = \dfrac{18}{(0.915) \ (8.34)}$

$x = 2.4 \text{ mg/L}$

323. $\dfrac{32 \text{ lb/day}}{0.60} = 53.3 \text{ lb/day Hypochlorite}$

324. $\dfrac{2,666,000 \text{ gal}}{7 \text{ days}} = 380,857 \text{ gpd } 4 \text{ in/12 in/ft} = 0.33 \text{ ft}$

$\dfrac{(0.785) \ (4 \text{ ft}) \ (4 \text{ ft}) \ (3.33 \text{ ft}) \ (7.48 \text{ gal/cu ft})}{7 \text{ days}} = 45 \text{ gpd}$

$(x \text{ mg/L}) \ (0.380 \text{ MGD}) \ (8.34 \text{ lb/gal}) = (20,000 \text{ mg/L}) \ (0.000026) \ (8.34)$

$x = \dfrac{(20,000) \ (0.000026) \ (8.34)}{(0.380) \ (8.34)}$

$x = 1.37 \text{ mg/L}$

325. Chlorine dose = 3.0 mg/L

(3.0 mg/L) (3.35 MGD) (8.34 lb/gal) = 83.8 lb/day

326. $\dfrac{(12 \text{ gal}) \ (12/100) \ + \ (50 \text{ gal}) \ (1/100)}{12 \text{ gal} + 50 \text{ gal}} \times 100$ $\dfrac{1.44 \text{ gal} + 0.5 \text{ gal}}{62 \text{ gal}} \times 100$

$\dfrac{1.94 \text{ gal}}{62 \text{ gal}} \times 100$

$= 3.1\%$

327. 72 lb/day = x/0.65

(72) (0.65) = x

46.8 lb/day = x

$(x \text{ mg/L}) \ (1.88 \text{ MGD}) \ (8.34 \text{ lb/gal}) = 46.8 \text{ lb/day}$

$x = \dfrac{46.8}{(1.88 \text{ MGD}) \ (8.34)}$

$x = 2.98 \text{ mg/L}$

328. (400 gpm) (1,440 min/day) = 576,000 gpd or 0.576 MGD

(2.6 mg/L) (0.576 MGD) (8.34 lb/gal) = (30,000 mg/L) (x MGD) (8.34 lb/gal)

$$\frac{(2.6)\ (0.576)\ (8.34)}{(30,000)\ (8.34)} = x$$

0.0000499 MGD = x

x = 49.9 gpd

$$\frac{(0.785)\ (3\ ft)^2\ (3\ ft)^2\ (4.08\ ft)\ (7.48\ gal/cu\ ft)^9}{92\ gpd} = 2.3\ days$$

330. $2 = \dfrac{x\ lb}{(80\ gal)\ (8.34\ lb/gal)\ +\ x\ lb} \times 100\ \ 2 = \dfrac{100\ x}{667.2\ +\ x}$

$$667.2 + x = \frac{100\ x}{2}$$

$667.2 + x = 50x$

$667.2 = 49x$

$$\frac{667.2}{49} = x$$

x = 13.7 lb

$$= \frac{13.7\ lb}{0.65}$$

= 21.1 lb hypochlorite

331. $\dfrac{32\ lb/day}{24\ hr/day} = 1.33$ lb/hr chlorine used 140 hr of operation

(1.33 lb/hr) (140 hr) = 186.2 lb

332. (50 lb/day) (30 days) = 1,500 lb

$$\frac{1500\ lb}{150\ lb/cylinder} = 10\ \text{cylinders required}$$

333. 2.6% = 26,000 mg/L

334. 6,600 mg/L = 0.67%

335. 29% = 290,000 mg/L

336. $\dfrac{22\ lb}{(1\ MG)\ (8.34\ lb/gal)} = \dfrac{22\ lb}{8.34\ mil\ lb} = \dfrac{2.64\ lb}{1\ mil\ lb} = 2.64$ mg/L

337. 1.6 mg/L $= \dfrac{1.6\ lb}{1\ mil\ lb} = \dfrac{1.6\ lb}{\dfrac{1\ mil\ lb}{8.34\ lb/gal}} = \dfrac{1.6\ lb}{0.12\ MG} = \dfrac{13.3}{1\ MG}$

338. $\dfrac{25\ lb}{(1\ MG)\ (8.34\ lb/gal)} = \dfrac{25\ lb}{8.34\ mil\ lb} = \dfrac{2.99\ lb}{1\ mil\ lb} = 2.99$ mg/L

339.

Element	Atoms		Atomic Wt	Molecular Wt
H	2	x	1.008	2.016
Si	1	x	28.06	28.06
F	6	x	19.00	114.00
				144.076

$$\frac{114.00}{144.076} \times 100$$

$$= 79.1\%$$

340.

Element	Atoms	Atomic Wt	Molecular Wt
Na	1	22.997	22.997
F	1	19.00	19.00
		Molecular wt of NaF	41.997

$$\frac{19.00}{41.997} \times 100 = 45.2\%$$

341.
$$\frac{(1.6 \text{ mg/L}) \ (0.98 \text{ MG}) \ (8.34 \text{ lb/gal})}{\dfrac{98}{100} \dfrac{(60.6)}{100}}$$

$$\frac{13.1}{(.98) \ (0.606)} = \frac{13.1}{0.59} = 22.2 \text{ lb/day Na}_2\text{SiF}_6$$

342.
$$\frac{(1.4 \text{ mg/L}) \ (1.78 \text{ MGD}) \ (8.34 \text{ lb/gal}}{(98/100) \qquad (60.6/100)} = 35.2 \text{ lb/day Na}_2\text{SiF}_6 \ (0.98)$$

$$(0.606) = 0.59$$

343.
$$\frac{(1.4 \text{ mg/L}) \ (2.880 \text{ MGD}) \ (8.34 \text{ lb/gal})}{0.8 \text{ lb}} = 42.1 \text{ lb/day Na}_2\text{SiF}_6$$

344.
$$\frac{(1.1 \text{ mg/L}) \ (3.08 \text{ MGD}) \ (8.34 \text{ lb/gal})}{0.45 \text{ lb}} = 62.8 \text{ lb/day NaF}$$

345. $1.2 \text{ mg/L} - 0.08 \text{ mg/L} = 1.13$
$$\frac{(1.13 \text{ mg/L}) \ (0.810 \text{ MGD}) \ (8.34 \text{ lb/gal})}{0.45 \text{ lb}} = 169 \text{ lb/day NaF}$$

346.
$$\frac{(91 \text{ lb}) \ (98/100)}{(55 \text{ gal}) \ (8.34 \text{ lb/gal}) + (9 \text{ lb}) \ (98/100)} \times 100 \quad \frac{8.82}{459 + 8.82} \times 100$$

$$= 1.9\% \text{ strength NaF}$$

347.
$$\frac{20 \text{ lb}}{(80 \text{ gal}) \ (8.34 \text{ lb/gal}) + (20 \text{ lb})} \times 100 \quad \frac{20 \text{ lb}}{667.2 + 20 \text{ lb}} \times 100$$

$$\frac{20 \text{ lb}}{687.2} \times 100$$

$$= 2.9 \text{ NaF}$$

348. $1.4 = \dfrac{(x \text{ lb}) \ (98/100)}{220 \text{ gal}) \ (8.34 \text{ lb/gal}) \ + \ (x \text{ lb}) \ (98/100)} \times 100$

$1.4 = \dfrac{0.98x}{1835 + 0.98 \text{ x}} \times 100$

$1.4 = \dfrac{98x}{1835 + 0.98 \text{ x}}$

$1.4 \ (1{,}835 + 0.98x) = 98x$

$2{,}569 + 137x = 98x$

$2{,}569 + 96.63x$

$x = 26.6 \text{ lb NaF}$

349. $\dfrac{(11 \text{ lb}) \ (98/100)}{(60 \text{ gal}) \ (8.34 \text{ lb/gal}) \ + \ (11 \text{ lb}) \ (98/100)} \times 100 = \dfrac{10.78}{500 + 10.8} \times 100$

$= 2.1\% \text{ strength NaF}$

350. $3 = \dfrac{(x \text{ lb}) \ (98/100)}{(160 \text{ gal}) \ (8.34 \text{ lb/gal}) + (\text{ x lb}) \ (98/100)} \times 100$

$3 = \dfrac{0.98x}{1334 + 0.98x} \times 100$

$3 = \dfrac{98x}{1334 + 0.98x}$

$3(1{,}334 + 0.98x) = 98x$

$1{,}334 + 2.94x = 98x$

$1{,}334 = 98x - 2.94x$

$1{,}334 = 95.06x$

$x = 14 \text{ lb NaF}$

351. $(1.2 \text{ mg/L}) \ (4.23 \text{ MGD}) \ (8.34 \text{ lb/gal}) = (240{,}000 \text{ mg/L}) \ (x) \ (8.34 \text{ lb/gal}) \ (1.2) \ (80/100)$

$\dfrac{(1.2) \ (4.23) \ (8.34)}{(240{,}000) \ (8.34) \ (1.2) \ (0.800)} = x \text{ MGD}$

$x = 0.0000217 \text{ MGD}$

$x = 21.7 \text{ gpd}$

352. $(1.2 \text{ mg/L}) \ (3.1 \text{ MGD}) \ (8.34 \text{ lb/gal}) = (220{,}000 \text{ mg/L}) \ (x) \ (9.7 \text{ lb/gal}) \ (80/100)$

$\dfrac{(1.2) \ (4.23) \ (8.34)}{(220{,}000) \ (9.7) \ (0800)} = x \text{ MGD} \quad 0.0000199 = x$

$x = 19.9 \text{ gpd}$

353. $1.8 \text{ mg/L} - 0.09 \text{ mg/L} = 1.71 \text{ mg/L}$

$(1.71 \text{ mg/L}) \ (0.91 \text{ MGD}) \ (8.34 \text{ lb/gal}) = (22{,}000 \text{ mg/L}) \ (x) \ (8.34 \text{ lb/gal}) \ (46.10/100)$

$\dfrac{(1.71) \ (0.91) \ (8.34)}{(22{,}000) \ (8.34) \ (0.4610)} = x \text{ MGD}$

$0.0001543 \text{ MGD} = x \text{ or } 154.3 \text{ MGD}$

354. (1.6 mg/L) (1.52 MGD) (8.34 lb/gal) = (24,000 mg/L) (x) (8.34 lb/gal) (45.25/100)

$$\frac{(1.6)\ (1.52)\ (8.34)}{(24,000)\ (8.34)\ (0.4575)} = x\,\text{MGD}$$

0.0002239 MGD = x

223.9 gpd = x

355. $\dfrac{(80\ \text{gpd})\ (3785\ \text{mL/gal})}{1440\ \text{min/day}} = 210.3\ \text{mL/min}$

356. (1.0 mg/L) (2.78 MGD) (8.34) = (250,000 mg/L) (x MGD) (9.8 lb/gal) (80/100)

$$\frac{(1.0)\ (2.78)\ (8.34)}{(250,000)\ (9.8)\ (0.80)} = x\,\text{MGD}$$

0.0000118 MGD = x

11.8 gpd = x

$$\frac{(11.8\ \text{gpd})\ (3785\ \text{mL/gal})}{1440\ \text{min/day}} = 31\ \text{mL/min}$$

357. $\dfrac{(x\ \text{mg/L})\ (1.52\ \text{MGD})\ (8.34\ \text{lb/gal})}{(98/100)\ (61/100)} = 40\ \text{lb/day}$ (x) (1.52) (8.34) = (40)

(0.98) (0.61)

$$x = \frac{(40)\ (0.98)\ (0.61)}{(1.52)\ (8.34)}$$

$$x = \frac{23.9}{12.7}$$

$x = 1.89$ mg/L F

358. $\dfrac{(x\ \text{mg/L})\ (0.33\ \text{MGD})\ (8.34\ \text{lb/gal})}{(98/100)\ (45.25)/100} = 6\ \text{lb/day}$ (x) (0.33) (8.34) = (6)

(0.98) (0.4525)

$$x = \frac{(6)\ (0.98)\ (0.4525)}{(0.33)\ (8.34)}$$

$x = 0.97$ mg/L F

359. (x mg/L) (3.85 MGD) (8.34 lb/gal) = (200,000 mg/L) (0.000032) (9.8 lb/gal) (80/100)

$$x = \frac{(200,000)\ (0.000032)\ (9.8)\ (0.80)}{(3.85)\ (8.34)}$$

$x = 1.6$ mg/L F

360. (x) (1.92 MGD) (8.34 lb/gal) = (110,000 mg/L) (0.000028 MGD) (9.10 lb/gal) (80/100)

$$x = \frac{(110{,}000)\ (0.000028)\ (9.10)\ (0.80)}{(1.92)\ (8.34)}$$

$x = 1.4$ mg/L F

361. $(x)\ (2.73\ \text{MGD})\ (8.34\ \text{lb/gal}) = (30{,}000\ \text{mg/L})\ (0.000110)\ (8.34\ \text{lb/gal})\ (45.25/100)$

$$x = \frac{(30{,}000)\ (0.000110)\ (8.34)\ (0.4525)}{(2.73)\ (8.34)}$$

$x = 0.55$ mg/L F

362. $\dfrac{(600\ \text{lb})\ (15/100) + (2600\ \text{lb})\ (25/100)}{600\ \text{lb} + 2600\ \text{lb}} \times 100 \quad \dfrac{90\ \text{lbs} + 650\ \text{lb}}{3200\ \text{lb}} \times 100$

$= 23\%$

363. $(900\ \text{lb})\ (25/100) + (300\ \text{lb})\ (15/100) = (1{,}200\ \text{lb})\ (x/100)$

$225\ \text{lb} + 45\ \text{lb} = (1{,}200)(x/100)$

$270 = 12x$

$270/12 = x$

$x = 22.5$

364. $\dfrac{(400\ \text{gal})\ (9.4\ \text{lb/gal})\ (16/100) + (2200)\ (9.10\ \text{lb/gal})\ (26/100)}{(400\ \text{gal})\ (19.4\ \text{lb/gal}) + (2200)\ (9.10\ \text{lb/gal})} \times 100$

$= \dfrac{601.6\ \text{lb} + 4404.4\ \text{lb}}{3760\ \text{lb} + 20{,}020\ \text{lb}} \times 100$

$\dfrac{5006}{23{,}780} = 21.1\%$

365. $\dfrac{(325\ \text{gal})\ (9.06\ \text{lb/gal})\ (11/100) + (1100\ \text{gal})\ (9.8\ \text{lb/gal})\ (20/100)}{(325\ \text{gal})\ (9.06\ \text{lb/gal}) + (1100)\ (9.8\ \text{lb/gal})} \times 100$

$= \dfrac{324\ \text{lb} + 2156}{2944.5\ \text{lb} + 10{,}780\ \text{lb}} \times 100$

$= \dfrac{2480\ \text{lb}}{13{,}724.5\ \text{lb}} \times 100$

$= 18.1\%$

366. Density $= (8.34\ \text{lb/gal})\ (1.075) = 8.97$

$\dfrac{(220\ \text{gal})\ (8.97\ \text{lb/gal})\ (10/100) + (1600)\ (9.5\ \text{lb/gal})\ (15/100)}{(220\ \text{gal})\ (8.97\ \text{lb/gal}) + (1600)\ (9.5\ \text{lb/gal})} \times 100$

$= \dfrac{197.3\ \text{lb} + 2280\ \text{lb}}{1973\ \text{lb} + 15{,}200\ \text{lb}} \times 100$

$= \dfrac{2477.3\ \text{lb}}{17173\ \text{lb}} \times 100$

$= 14.4\%$

367. $2.9000 = 29{,}000$ mg/L

368. Molecular wt $= 2 \times 1.008 = 2.016$

$1 \times 28.06 = 28.06$

$6 \times 19.00 = 114.00$

144.076

$$= \frac{114.00}{144.076} \times 100$$

$= 79.1\%$

369. $$= \frac{27 \text{ lb}}{(1 \text{ MG}) \ (8.34 \text{ lb/gal})} = \frac{27 \text{ lb}}{8.34 \text{ mil lb}} \times \frac{3.24}{1 \text{ mil lb}} = 3.24 \text{ mg/L}$$

370. $$\frac{(1.6 \text{ mg/L}) \ (2.111 \text{ MGD}) \ (8.34 \text{ lb/gal})}{(98/100) \ (61.2/100)} = 47 \text{ lb/day}$$

371. Molecular wt of NaF = 41.997

$$= \frac{19.00}{41.997} \times 100$$

$= 45.2\%$

372. $$\frac{(80 \text{ lb}) \ (98/100)}{(600 \text{ gal}) \ (8.34 \text{ lb/gal}) \ + \ (80 \text{ lb}) \ (98/100)} \times 100$$

$$= \frac{78.4 \text{ lb}}{5004 \text{ lb} + 78.4 \text{ lb}} \times 100$$

$$= \frac{78.4}{5082.4} \times 100$$

$= 1.5\%$

373. 28,000 mg/L = 2.8%

374. $$\frac{(80 \text{ gpd}) \ (3785 \text{ mL/gal})}{1440 \text{ min/day}} = 210 \text{ mL/min}$$

375. $$\frac{(1.5 \text{ mg/L}) \ (2.45 \text{ MGD}) \ (8.34 \text{ lb/gal})}{(98/100) \ (45.25/100)} = 156 \text{ lb/day}$$

376. $$3 = \frac{(x \text{ lb}) \ (98/100)}{(600 \text{ gal}) \ (8.34 \text{ lb/gal}) \ + \ (x \text{ lb}) \ (98/100)} \times 100$$

$$3 = \frac{0.98 \ x}{5004 \text{ lb} + 0.98 \ x} \times 100$$

$$3 = \frac{98x}{5004 \text{ lb} + 0.98 \ x}$$

$3(5,004 + 0.98x) = 98x$

$15,012 + 2.94x = 98x$

$15,012 = 95.06x$

$x = 158 \text{ lb NaF}$

377. (1.4 mg/L) (4.11) (8.34 lb/gal) = (210,000 mg/L) (x MGD) (8.34 lb/gal) (1.3) (80/100)

$$\frac{(1.4)\ (4.11)\ (8.34)}{(210,00)\ (8.34)\ (1.3)\ (0.8)} = x \text{ MGD}$$

$0.0000262 \text{ MGD} = x$

$26.2 \text{ gpd} = x$

378. $\dfrac{(30 \text{ lb})\ (98/100)}{(140 \text{ gal})\ (8.34 \text{ lb/gal})\ +\ (30 \text{ lb})\ (98/100)} \times 100$

$= \dfrac{29.4 \text{ lb}}{1167.8 \text{ lb} + 29.4} \times 100$

$= \dfrac{29.4}{1197.2} \times 100$

$= 2.45\%$

379. $1.4 \text{ mg/L} - 0.09 \text{ mg/L} = 1.31 \text{ mg/L}$

$\dfrac{(1.31 \text{ mg/L})\ (1.88 \text{ MGD})\ (8.34 \text{ lb/gal})}{0.44} = 46.7 \text{ lb/day NaF}$

380. $(1.3 \text{ mg/L})\ (2.8 \text{ MGD})\ (8.34 \text{ lb/gal}) = (200,000)\ (x \text{ MGD})\ (9.8 \text{ lb/gal})$
$(80/100)$

$\dfrac{(1.3)\ (2.8)\ (8.34)}{(200,000)\ (9.8)\ (0.80)} \times \text{ MGD}$

$0.0000193 \text{ MGD} = x$

$19.3 \text{ gpd} = x$

381. $\dfrac{(500 \text{ lb})\ (15/100)\ +\ (1600 \text{ lb})\ (20/100)}{500 + 1600 \text{ lb}} \times 100$

$= \dfrac{75 \text{ lbs} + 320 \text{ lb}}{2100 \text{ lb}} \times 100$

$= \dfrac{395 \text{ lb}}{2100 \text{ lb}} \times 100$

$= 18.8\%$

382. $\dfrac{(x \text{ mg/L})\ (1.10 \text{ MGD})\ (8.34 \text{ lb/gal})}{(98/100)\qquad (61.1/100)} = 41 \text{ lb/day } (x)\ (1.10)\ (8.34) = (41)$

$(0.98)\ (0.611)$

$x = \dfrac{(41)\ (0.98)\ (0.611)}{(1.10)\ (8.34)}$

$x = 2.7 \text{ mg/L F}$

383. $(1,400 \text{ gpm})\ (1,440 \text{ min/day}) = 2,016,000 \text{ gpd} = 2.016 \text{ MGD}$
$(x \text{ mg/L})\ (2.016 \text{ flow})\ (8.34 \text{ lb/gal}) = (110,000 \text{ mg/L})\ (0000.40 \text{ MGD})\ (9.14$
$\text{lb/gal})$

$x = \dfrac{(110,000)\ (0.000040)\ (9.14)\ (0.8)}{(2.016)\ (8.34)}$

$x = 1.92 \text{ mg/L}$

384. $\dfrac{(235 \text{ gal}) \ (9.14 \text{ lb/gal}) \ (10/100) \ + \ (600 \text{ gal}) \ (9.8 \text{ lb/gal}) \ (20/100)}{(235 \text{ gal}) \ (9.14 \text{ lb/gal}) \ + \ (600 \text{ gal}) \ (9.8 \text{ lb/gal})} \times 100$

$= \dfrac{215 \text{ lb} + 1176 \text{ lb}}{2147.9 \text{ lb} + 5880 \text{ lb}} \times 100$

$= \dfrac{1391}{8027.9} \times 100$

$= 17.3\%$

385. $(1.1 \text{ mg/L}) \ (2.88 \text{ MGD}) \ (8.34 \text{ lb/gal}) = (200{,}000 \text{ mg/L}) \ (x \text{ MGD}) \ (9.8 \text{ lb/gal})$ $(80/100)$

$\dfrac{(1.1) \ (2.88) \ (8.34)}{(200{,}000) \ (9.8) \ (0.8)} \times \text{MGD}$

$0.0000168 \text{ MGD} = x$

$16.8 \text{ gpd} = x$

$\dfrac{(16.8 \text{ gpd}) \ (3785 \text{ mL/gal})}{1440 \text{ min/day}} = 44.2 \text{ mL/min}$

386. Density $= (8.34 \text{ lb/gal}) \ (1.115) = 9.3 \text{ lb/gal}$

$\dfrac{(131 \text{ gal}) \ (9.3 \text{ lb/gal}) \ (9/100) + 900 \text{ gal}) \ (9.4 \text{ lb/gal}) \ (15/100)}{(131 \text{ gal}) \ (9.3 \text{ lb/gal}) + (900 \text{ gal}) \ (9.4 \text{ lb/gal})} \times 100$

$\dfrac{(109.6) + (1269 \text{ lb})}{(1218 \text{ lb}) + (8460 \text{ lb})} \times 100$

$\dfrac{1378.6}{9678} \times 100$

$= 14.2\%$

387. $(x \text{ mg/L}) \ (2.90) \ (8.34 \text{ lb/gal}) = (50{,}000 \text{ mg/L}) \ (0.000120) \ (8.34 \text{ lb/gal})$ $(45.25/100)$

$x = \dfrac{(50{,}000) \ (0.000120) \ (8.34) \ (0.4525)}{(2.90 \text{ MGD}) \ (8.34)}$

$x = 0.94 \text{ mg/L F}$

$0.94 \text{ mg/L F added} + 0.2 \text{ mg/L in raw water} = 0.96 \text{ in finished water}$

388. $\dfrac{x \text{ m/L}}{50.045} = \dfrac{39 \text{ mg/L}}{20.04} \quad x = \dfrac{(39) \ (50.045)}{20.04}$

$x = 98.2 \text{ mg/L calcium as } CaCO_3$

389. $\dfrac{x \text{ mg/L}}{50.045} = \dfrac{33 \text{ mg/L}}{12.16} \quad x = \dfrac{(33) \ (50.045)}{12.16}$

$x = 136 \text{ mg/L magnesium as } CaCO_3$

390. $\dfrac{x \text{ mg/L}}{50.045} = \dfrac{18 \text{ mg/L}}{20.04} \quad x = \dfrac{(18) \ (50.045)}{20.04}$

$x = 44.9 \text{ mg/L calcium as } CaCO_3$

391. $75 \text{ mg/L} + 91 \text{ mg/L} = 166 \text{ mg/L as } CaCO_3$

393. $\dfrac{x \text{ mg/L}}{50.045} = \dfrac{30 \text{ mg/L}}{20.04} \quad x = \dfrac{(30)\,(50.045)}{20.04}$

$x = 74.9$ mg/L calcium as $CaCO_3$

$\dfrac{x \text{ mg/L}}{50.045} = \dfrac{10 \text{ mg/L}}{12.16}$

$x = \dfrac{(10)\,(50045)}{12.16}$

$x = 41$ mg/L magnesium as $CaCO_3$

Total hardness mg/L as $CaCO_3 = 74.9$ mg/L $+ 41$ mg/L
$= 115.9$ mg/L total hardness

393. $\dfrac{x \text{ mg/L}}{50.045} = \dfrac{21 \text{ mg/L}}{20.04}$

$x = \dfrac{(21)\,(50.045)}{20.04}$

$x = 52.4$ mg/L calcium as $CaCO_3$

$\dfrac{x \text{ mg/L}}{50.045} = \dfrac{15 \text{ mg/L}}{12.16}$

$x = \dfrac{(15)\,(50.045)}{12.16}$

$x = 61.7$ mg/L magnesium as $CaCO_3$
$= 52.4$ mg/L $+ 61.7$ mg/L
$= 114.1$ mg/L total hardness as $CaCO_3$

394. Total hardness, mg/L as $CaCO_3 =$ carbonate hardness, mg/L as 121 mg/L as $CaCO_3 =$ carbonate hardness. There is no noncarbonate hardness in this water.

395. 122 mg/L = 105 mg/L + x mg/L
122 mg/L − 105 mg/L = x mg/L
Noncarbonate hardness 17 mg/L = x
Carbonate hardness is 105 mg/L

396. 116 mg/L = 91 mg/L + x mg/L
116 mg/L − 91 mg/L = x mg/L
Noncarbonate hardness 25 mg/L = x
Carbonate hardness is 91 mg/L.

397. Alkalinity is greater than total hardness. Therefore, all the hardness is carbonate hardness.
99 mg/L as $CaCO_3 =$ carbonate hardness

398. 121 mg/L = 103 mg/L + mg/L
121 mg/L − 103 mg/L = x mg/L
Noncarbonate hardness 18 mg/L = x
Carbonate hardness is 103 mg/L.

399. Phenolphthalein Alk, mg/L as $CaCO_3 = \dfrac{(A)\,(N)\,(50,000)}{\text{ml of sample}}$

$$= \frac{(2.0 \text{ mL}) \ (0.02\text{N}) \ (50,000)}{100 \text{ mL}}$$

$= 20$ mg/L as $CaCO_3$ phenolphthalein alkalinity

400. $= \frac{(1.4 \text{ mL}) \ (0.02\text{N}) \ (50,000)}{100 \text{ mL}}$

$= 14$ mg/L

401. $\frac{(0.3 \text{ mL}) \ (0.02\text{N}) \ (50,000)}{100 \text{ mL}} = 3$ mL

Total Alkalinity $= \dfrac{(6.7 \text{ mL}) \ (0.02\text{N}) \ (50,000)}{100 \text{ mL}}$

$= 67$ mg/L

402. Phenolphthalein alkalinity $= 0$ mg/L

Total Alkalinity as $CaCO_3 = \dfrac{(6.9 \text{ mL}) \ (0.02\text{N}) \ (50,000)}{100 \text{ mL}}$

$= 69$ mg/L

403. Phenolphthalein alk mg/L as $CaCO_3 =$

$\dfrac{(0.5 \text{ mL}) \ (0.02\text{N}) \ (50,000)}{100 \text{ mL}} = 5$ mg/L

Total Alkalinity $= \dfrac{(5.7 \text{ mL}) \ (0.02\text{N}) \ (50,000)}{100 \text{ mL}} = 57$ mg/L

404. Bicarbonate alkalinity $= T - 2P$
   $= 51$ mg/L $- 2(8$ mg/L$)$
   $= 51$ mg/L $- 16$ mg/L
   $= 35$ mg/L as $CaCO_3$
   Carbonate alkalinity $= 2P$
   $= 2(8$ mg/L$)$
   $= 16$ mg/L as $CaCO_3$
   Hydroxide alkalinity $= 0$ mg/L

405. Bicarbonate alkalinity $= T$
   $= 67$ mg/L as $CaCO_3$
   Carbonate alkalinity $= 0$ mg/L
   Hydroxide alkalinity $= 0$ mg/L

406. Bicarbonate alkalinity $= 0$ mg/L
   Carbonate alkalinity $= 2T - 2P$
   $= 2(23$ mg/L$) - 2(12$ mg/L$)$
   $= 46$ mg/L $- 24$ mg/L
   $= 22$ mg/L as $CaCO_3$
   Hydroxide alkalinity $= 2P - T$
   $= 2(12$ mg/L$) - 23$ mg/L
   $= 24$ mg/L $- 23$ mg/L
   $= 1$ mg/L as $CaCO_3$

407. Phenolphthalein Alk, mg/L as $CaCO_3 = \dfrac{(1.3 \text{ mL}) \ (0.02\text{N}) \ (50{,}000)}{100 \text{ mL}}$

= 13 mg/L as $CaCO_3$

Total Alk., mg/L as $CaCO_3 = \dfrac{(5.3 \text{ mL}) \ (0.02\text{N}) \ (50{,}000)}{100 \text{ mL}}$

= 53 mg/L as $CaCO_3$
Bicarbonate alk = T – 2P
= 51 mg/L – 2(13 mg/L)
= 51 mg/L – 26 mg/L
= 25 mg/L as $CaCO_3$
Carbonate alk = 2P
= 2 (13 mg/L)
= 26 mg/L as $CaCO_3$
Hydroxide alk = 0 mg/L

408. Phenolphthalein Alkalinity, mg/L as $CaCO_3 = \dfrac{(1.5 \text{ mL}) \ (0.02\text{N}) \ (50{,}000)}{100 \text{ mL}}$

= 15 mg/L as $CaCO_3$

Total Alk., mg/L $CaCO_3 = \dfrac{(2.9 \text{ mL}) \ (0.02\text{N}) \ (50{,}000)}{100 \text{ mL}}$

= 29 mg/L as $CaCO_3$
From the alkalinity table:
Bicarbonate alkalinity = 0 mg/L
Carbonate alkalinity = 2P
= 2(15 mg/L)
= 30 mg/L as $CaCO_3$
Hydroxide alkalinity = 0 mg/L

409. A = (CO_2, mg/L) (56/44)
= (8 mg/L) (56/44)
= 10 mg/L
B = (alkalinity, mg/L) (56/100)
= (130 mg/L) (56/100)
= 73 mg/L
C = 0 mg/L
D = (Mg^{+2}, mg/L) (56/24.3)
= (22 mg/L) (56/24.3)
= 51 mg/L

Quicklime Dosage, mg/L $= \dfrac{(10 \text{ mg/L}) + (73 \text{ mg/L}) + 0 + (51 \text{ mg/L}) \ (1.15)}{0.90}$

$= \dfrac{(134 \text{ mg/L}) \ (1.15)}{0.90}$

= 171 mg/L CaO

410. A = (CO_2, mg/L) (74/44)
= (5 mg/L) (74/44)

= 8 mg/L

B = (alkalinity, mg/L) (74/100)

= (164 mg/L) (74/100)

= 121 mg/L

$C = 0$ mg/L

$D = (Mg^{+2}, mg/L) (74/24.3)$

= (17 mg/L) (74/24.3)

= 52 mg/L

$$\text{Hydrated Lime Dosage, mg/L} = \frac{\left(8\ mg/L + 121\ mg/L + 0 + 52\ mg/L\right)\ (1.15)}{0.90}$$

$$= \frac{(181\ mg/L)\ (1.15)}{0.90}$$

= 231 mg/L Ca(OH)$_2$

411. $A = (CO_2, mg/L) (74/44)$

= (6 mg/L) (74/44)

= 10 mg/L

B = (alkalinity, mg/L) (24/100)

= (110 mg/L) (74/100)

= 81 mg/L

$C = 0$ mg/L

$D = (Mg^{+2}, mg/L) (74/24.3)$

= (12 mg/L) (74/24.3)

= 37 mg/L

$$\text{Hydrated Lime} = \frac{\left(10\ mg/L + 81\ mg/L + 0 + 37\ mg/L\right)\ (1.15)}{0.90}$$

$$= \frac{(128\ mg/L)\ 1.15}{0.90}$$

= 164 mg/L Ca(OH)$_2$

412. A = (carbon dioxide, mg/L) (56/44)

= (9 mg/L) (56/44)

= 11 mg/L

B = (alkalinity, mg/L) (56/100)

= (180 mg/L) (56/100)

= 101 mg/L

$C = 0$ mg/L

$D = (Mg^{+2}, mg/L) (56/24.3)$

= (18 mg/L) (56/24.3)

= 41 mg/L

$$\text{Quicklime Dosage, mg/L} = \frac{\left(11\ mg/L + 101\ mg/L + 0 + 41\ mg/L\right)\ (1.15)}{0.90}$$

$$= \frac{(153\ mg/L)\ (1.15)}{0.90}$$

= 196 mg/L CaO

413. $260 \text{ mg/L} = 169 \text{ mg/L} = x \text{ mg/L}$
$260 \text{ mg/L} - 169 \text{ mg/L} = x$
$91 \text{ mg/L} = x$
Soda ash, mg/L = $(91 \text{ mg/L}) (106/100)$
$= 96 \text{ mg/L soda ash}$

414. Noncarbonate hardness
$240 \text{ mg/L} = 111 \text{ mg/L} + x \text{ mg/L}$
$240 \text{ mg/L} - 111 \text{ mg/L} = x$
$129 \text{ mg/L} = x$
Soda ash, mg/L = $(129 \text{ mg/L}) (106/100)$
$= 137 \text{ mg/L soda ash}$

415. Noncarbonate hardness
$264 \text{ mg/L} = 170 \text{ mg/L} + x \text{ mg/L}$
$264 \text{ mg/L} - 170 \text{ mg/L} = x$
$94 \text{ mg/L} = x$
Soda ash, mg/L = $(94 \text{ mg/L}) (106/100)$
$= 100 \text{ mg/L soda ash}$

416. $228 \text{ mg/L} = 108 \text{ mg/L} + x \text{ mg/L}$
$228 \text{ mg/L} - 108 \text{ mg/L} = x$
$120 \text{ mg/L} = x$
Soda ash, mg/L = $(120 \text{ mg/L}) (106/100)$
$= 127 \text{ mg/L soda ash}$

417. Excess lime, mg/L = $(8 \text{ mg/L} + 130 \text{ mg/L} + 0 + 66 \text{ mg/L}) (0.15)$
$= (204 \text{ mg/L}) (0.15)$
$= 31 \text{ mg/L}$
Total carbon dioxide dosage, mg/L = $(31 \text{ mg/L}) (44/100) + (4 \text{ mg/L}) (44/24.3)$
$= (14 \text{ mg/L}) + 7 \text{ mg/L}$
$= 21 \text{ mg/L carbon dioxide}$

418. Excess lime, mg/L = $(8 \text{ mg/L} + 90 \text{ mg/L} + 7 + 109 \text{ mg/L}) (0.15)$
$= (213 \text{ mg/L}) (0.15)$
$= 32 \text{ mg/L}$
Total carbon dioxide, mg/L = $(32 \text{ mg/L}) (44/74) + (3 \text{ mg/L}) (44/24.3)$
$= 19 \text{ mg/L} + 5.4 \text{ mg/L}$
$= 24.4 \text{ mg/L}$

419. Excess lime, mg/L = $(7 \text{ mg/L} + 109 \text{ mg/L} + 3 + 52 \text{ mg/L}) (0.15)$
$= (171 \text{ mg/L}) (0.15)$
$= 26 \text{ mg/L}$
Total carbon dioxide, mg/L = $(26 \text{ mg/L}) (44/74) + (5 \text{ mg/L}) (44/24.3)$
$= 15.5 \text{ mg/L} + 9 \text{ mg/L}$
$= 24.5 \text{ mg/L carbon dioxide}$

420. Excess lime, mg/L = $(6 \text{ mg/L} + 112 \text{ mg/L} + 6 + 45 \text{ mg/L}) (0.15)$
$= (169 \text{ mg/L}) (0.15)$
$= 26 \text{ mg/L}$
Total carbon dioxide dosage, mg/L = $(26 \text{ mg/L}) (44/74) + (4 \text{ mg/L}) (44/24.3)$

= 15 mg/L + 7 mg/L

= 22 mg/L carbon dioxide

421. (200 mg/L) (2.47 MGD) (8.34 lb/gal) = 4,120 lb/day

422. (180 mg/L) (3.12 MGD) (8.34 lb/gal) = 4,684 lb/day

$$\frac{4684 \text{ lb/day}}{1440 \text{ min/day}} = 3.3 \text{ lb/min}$$

423. (60 mg/L) (4.20 MGD) (8.34 lb/gal) = 2,102 lb/day

$$\frac{2102 \text{ lb/day}}{24 \text{ hr/day}} = 87.5 \text{ lb/hr}$$

424. (130 mg/L) (1.85 MGD) (8.34 lb/gal) = 2,006 lb/day

$$\frac{2006 \text{ lb/day}}{1440 \text{ min/day}} = 1.4 \text{ lb/min}$$

425. (40 mg/L) (3.11 MGD) (8.34 lb/gal) = 1,038 lb/day

$$\frac{1038 \text{ lb/day}}{24 \text{ hr/day}} = 43 \text{ lb/hr}$$

$$\frac{43 \text{ lb/hr}}{60 \text{ min/hr}} = 0.7 \text{ lb/min}$$

426. (17.12 mg/L/gpg)

$$\frac{211 \text{ mg/L}}{17.12 \text{ mg/L/gpg}} = 12.3 \text{ gpg}$$

427. (17.12 mg/L/gpg)

(12.3 mg/L) (17.12 mg/L/gpg) = 211 mg/L

428. $\dfrac{240 \text{ mg/L}}{17.12 \text{ mg/L/gpd}} = 14 \text{ gpg}$

429. (14 gpg) (17.12 mg/L/gpg) = 240 mg/L

430. (25,000 grains/cu ft) (105 cu ft) = 2,625,000 grains

431. (0.785) (6 ft) (6 ft) (4.17 ft)

= 119 cu ft

= (25,000 grains/cu ft) (119 cu ft)

= 2,975,000 grains

432. (20,000 grains/cu ft) (260 cu ft) = 5,200,000 grains

433. (0.785) (8ft) (8 ft) (5 ft) = 251 cu ft

(22,000 grains/cu ft) (251 cu ft) = 5,522,000 grains

434. $\dfrac{2,210,000 \text{ grains}}{18.1 \text{ gpg}} = 122,099 \text{ gallons}$

435. $\dfrac{4,200,000 \text{ grains}}{16.0 \text{ gpg}} = 262,500 \text{ gallons}$

436. $\dfrac{270 \text{ mg/L}}{17.12 \text{ mg/L/gpg}} = 15.8 \text{ gpg}$ $\dfrac{3,650,000 \text{ grains}}{15.8 \text{ gpg}} = 231,013 \text{ gallons}$

437. (21,000 grains/cu ft) (165 cu ft) = 3,465,000 grains

$$\frac{3,465,000 \text{ grains}}{14.6 \text{ gpg}} = 237,329 \text{ gallons}$$

438. (0.785) (3 ft) (3 ft) (2.6 ft) = 18.4 cu ft

     (22,000 grains/cu ft) (18.4 cu ft) = 404,800 grains

$$\frac{404,800 \text{ grains}}{18.4 \text{ gpg}} = 22,000 \text{ gallons}$$

439. $\dfrac{575,000 \text{ gal}}{25,200 \text{ gph}} = 22.8 \text{ hr of operation}$

440. $\dfrac{766,000 \text{ gal}}{26,000 \text{ gph}} = 29.5 \text{ hr of operation}$

441. (230 gpm) (60 min/hr) = 13,800 gph

$$\frac{348,000 \text{ gal}}{13,800 \text{ gph}} = 25.2 \text{ hr of operating time}$$

442. $\dfrac{3,120,000 \text{ grains}}{14 \text{ gpg}} = 222,857 \text{ gallons}$ (200 gpm) (60 min/hr) = 12,000 gph

$$\frac{222,857 \text{ gal}}{12,000 \text{ gph}} = 18.6 \text{ hrs operating time}$$

443. $\dfrac{3,820,000 \text{ grains}}{11.6 \text{ gpg}} = 329,310 \text{ gallons}$ $\dfrac{290,000 \text{ gpd}}{24 \text{ hr/day}} = 12,083 \text{ gph}$

$$\frac{329,310 \text{ gal}}{12,083 \text{ gph}} = 27.3 \text{ hr operating}$$

444. $\dfrac{(0.5 \text{ lb salt})(2300 \text{ kilograins})}{\text{kilograins rem.}} = 1,150 \text{ lb salt required}$

445. $\dfrac{(0.4 \text{ lb salt})(1330 \text{ kilograins})}{\text{kilograins rem.}} = 532 \text{ lb salt required}$

446. $\dfrac{410 \text{ lb salt}}{1.19 \text{ lb salt/gal brine}} = 345 \text{ gallons of } 13\% \text{ brine}$

447. $\dfrac{420 \text{ lb salt}}{1.29 \text{ lb salt/gal brine}} = 326 \text{ gallons of } 14\% \text{ brine}$

448. $\dfrac{0.5 \text{ lb salt}}{\text{kilograins rem.}} (1310 \text{ kilograins}) = 655 \text{ lb salt required}$

$$\frac{655 \text{ lb salt}}{1.09 \text{ lb salt/gal brine}} = 601 \text{ gallons of } 12\% \text{ brine}$$

449. $\dfrac{x \text{ mg/L}}{50.045} = \dfrac{44 \text{ mg/L}}{20.04}$ $x \text{ mg/L} = \dfrac{(44 \text{ mg/L}) (50.045)}{20.04}$

     $x = 110 \text{ mg/L Ca as CaCO}_3$

450. $\dfrac{(1.8 \text{ mL}) (0.02N) (50,000)}{100 \text{ mL}} = 18 \text{ mg/L as CaCO}_3$

451. $\dfrac{x \text{ mg/L}}{50.045} = \dfrac{31 \text{ mg/L}}{12.16}$ $x = \dfrac{(31) (50.045)}{12.16} = 128 \text{ mg/L as CaCO}_3$

452. $\dfrac{24 \text{ mg}}{12.16} = 1.97$ milliequivalents

453. $A = (8 \text{ mg/L}) (74/44)$
     $= 13 \text{ mg/L}$
     $B = (118 \text{ mg/L}) (74/100)$
     $= 87 \text{ mg/L}$
     $C = 0$
     $D = (12 \text{ mg/L}) (74/24.3)$
     $= 37 \text{ mg/L}$
     $$\dfrac{(13 \text{ mg/L}) + (87 \text{ mg/L}) + 0 + (37 \text{ mg/L}) \ (1.15)}{.90}$$
     $$\dfrac{(137 \text{ mg/L}) \ (1.15)}{0.90} = 158 \text{ mg/L Ca(OH)}_2$$

454. Calcium hardness
     $$\dfrac{x \text{ mg/L}}{50.045} = \dfrac{31 \text{ mg/L}}{20.04}$$
     $x = 77 \text{ mg/L Ca as CaCO}_3$
     Magnesium hardness
     $$\dfrac{X \text{ mg/L}}{50.045} = \dfrac{11 \text{ mg/L}}{12.16}$$
     $x = 45 \text{ mg/L Mg as CaCO}_3$
     Total hardness, mg/L as $CaCO_3$
     $= 77 \text{ mg/L} + 45 \text{ mg/L}$
     $= 122 \text{ mg/L total hardness as CaCO}_3$

455. $101 \text{ mg/L as CaCO}_3$ = carbonate hardness

456. $A = (5 \text{ mg/L}) (56/44)$
     $= 6 \text{ mg/L}$
     $B = (156 \text{ mg/L}) (56/100)$
     $= 87 \text{ mg/L}$
     $C = 0 \text{ mg/L}$
     $D = (11 \text{ mg/L}) (56/24.3)$
     $= 25 \text{ mg/L}$
     Quicklime Dosage, mg/L $= \dfrac{(6 \text{ mg/L}) + (87 \text{ mg/L}) + 0 + (25 \text{ mg/L}) \ (1.15)}{0.90}$
     $$\dfrac{(118 \text{ mg/L}) \ (1.15)}{0.90} = 151 \text{ mg/L CaO}$$

Answers to Additional Practice Problems

1. $(0.785) (70 \text{ ft}) (70 \text{ ft}) (25 \text{ ft}) (7.48 \text{ gal/cu ft}) = 719,295.5 \text{ gal}$
2. $(60 \text{ ft}) (20 \text{ ft}) (10 \text{ ft}) = 12,000 \text{ cu ft}$

3. (20 ft) (60 ft) (12 ft) (7.48 gal/cu ft) = 107,712 gal
1. (20 ft) (40 ft) (12 ft) (7.48 cu ft) = 71,808 gal
2. (0.785) (60 ft) (60 ft) (12 ft) (7.48 gal/cu ft) = 253,662 gal
3. (20 ft) (50 ft) (16 ft) (7.48 gal/cu ft) = 119,680 gal
7. (4 ft) (6 ft) (340 ft) = 8,160 cu ft
8. (0.785) (0.83 ft) (0.83 ft) (1,600 ft) (7.48 gal/cu ft) = 6,472 gal
9. 5 ft + 10 ft/2 (4 ft) (800 ft) (7.48 gal/cu ft)
   = (7.5 ft) (4 ft) (800 ft) (7.48 gal/cu ft)
   = 179,520 gal
10. (0.785) (.66) (.66) (2,250 ft) (7.48 gal/cu ft) = 5,755 gal
11. (5 ft) (4 ft) (1,200 ft) (7.48 gal/cu ft) = 179,520 gal
12. $\dfrac{(4\ ft)(4\ ft)(1200\ ft)}{27\ cu\ ft/cu\ yd} = 711$ cu yds
13. (500 yd) (1 yd) (1.33 yd) = 665 cu yd
14. (900 ft) (3 ft) (3 ft) = 8,100 cu ft
15. (700 ft) (6.5 ft) (3.5 ft) = 15,925 cu ft
16. (0.785) (90 ft) (90 ft) (25 ft) (7.48 gal/cu ft) = 1,189,040 gal
17. (80 ft) (16 ft) (20 ft) = 25,600 cu ft
18. (0.785) (0.67 ft) (0.67 ft) (4,000 ft) (7.48 gal/cu ft) = 10,543 gal
19. (1,200 ft) (3 ft) (3 ft) = 10,800 cu ft
20. $\dfrac{(3\ ft)(4\ ft)(1200\ ft)}{27\ cu\ ft/cu\ yd} = 533$ cu yds
21. (30 ft) (80 ft) (12 ft) (7.48 gal/cu ft) = 215,424 gal
22. (8 ft) (3.5 ft) (3,000 ft) (7.48 gal/cu ft) = 628,320 gal
23. (0.785) (70 ft) (70 ft) (19 ft) (7.48 gal/cu ft) = 546,665 gal
24. (0.785) (25 ft) (25 ft) (30 ft) (7.48 gal/cu ft) = 110,096 gal
25. (2.4 ft) (3.7 ft) (2.5 fps) (60 sec/min) = 1,332 cfm
26. (20 ft) (12 ft) (0.8 fpm) (7.48 gal/cu ft) = 1,436 gpm
27. $\dfrac{(4\ ft+6\ ft)}{2}(3.3\ ft)\ (130\ fpm) = 5(3.3\ ft)\ (130\ fpm)$
    = 2,145 cfm
28. (0.785) (0.66) (0.66) (2.4 fps) (7.48 gal/cu ft) (60 sec/min) = 368 gpm
29. (0.785) (3 ft) (3 ft) (4.7 fpm) (7.48 gal/cu ft) = 248 gpm
30. (0.785) (0.83 ft) (0.83 ft) (3.1 fps) (7.48 gal/cu ft) (60 sec/min) (0.5) = 376 gpm
31. (6 ft) (2.6 ft) (x fps) (60 sec/min) (7.48 gal/cu ft) = 14,200 gpm
    $x = 2.03$ ft
32. (0.785) (0.67) (0.67) (x fps) (7.48 gal/cu ft) (60 sec/min) = 584 gpm
    $x = 3.7$ fps
33. 550 ft/208 sec = 2.6 fps
34. (0.785) (0.83 ft) (0.83 ft) (2.4 fps) = (0.785) (0.67 ft) (0.67 ft) (x fps)
    $x = 3.7$ fps
35. 500 ft/92 sec = 5.4 fps

36. (0.785) (0.67) (0.67) (3.2 fps) = (0.785) (0.83 ft) (0.83 ft) (x fps)
x = 2.1 fps

37. 35.3 MGD/7 = 5 MGD

38. 121.4 MG/30 days = 4.0 MGD

39. 1,000,000 x 0.165 = 165,000 gpd

40. 3,335,000 gal/1,440 min = 2,316 gpm

41. (8 cfs) (7.48 gal/cu ft) (60 sec/min) = 3,590 gpm

42. (35 gps) (60 sec/min) (1,440 min/day) = 3,024,000 gpd

43. $$\frac{4,570,000 \text{ gpd}}{(1440 \text{ min/day}) \ (7.48 \text{ gal/cu ft})} = 424 \text{ cfm}$$

44. (6.6 MGD) (1.55 cfs/MGD) = 10.2 cfs

45. $$\frac{(445,875 \text{ cfd})(7.48 \text{ gal / cu ft})}{1440 \text{ min/day}} = 2316 \text{ gpm}$$

46. (2,450 gpm) (1,440 min/day) = 3,528,000 gpd

47. (6 ft) (2.5 ft) (x fps) (7.48 gal/cu ft) (60 sec/min) = 14,800 gpm
x = 2.2 fps

48. (4.6 ft) (3.4 ft) (3.6 fps) (60 sec/min) = 3,378 cfm

49. 373.6/92 days = 4.1 MGD

50. (12 ft) (12 ft) (0.67 fpm) (7.48 gal/cu ft) = 722 gpm

51. (0.785) (0.67 ft) (0.67 ft) (x fps) (7.48 gal/cu ft) (60 sec/min) = 510 gpm
x = 3.2 fps

52. (10 cfs) (7.48 gal/cu ft) (60 sec/min) = 4,488 gpm

53. 134.6/31 days = 4.3 MGD

54. (5.2 MGD) (1.55 cfs/MGD) = 8.1 cfs

55. (0.785) (2 ft) (2 ft) (3.3 fpm) (7.48 gal/cu ft) = 77.5 gpm

56. $$\frac{(1,825,000 \text{ gpd})}{(1440 \text{ min/day}) \ (7.48 \text{ gal/cu ft})} = 169 \text{ cfm}$$

57. (0.785) (0.5 ft) (0.5 ft) (2.9 fps) (7.48 gal/cu ft) (60 sec/min) = 255 gpm

58. (0.785) (0.83 ft) (0.83 ft) (2.6 fps) = (0.785) (0.67 ft) (0.67 ft) (x fps) =
x = 4.0 fps

59. (2,225 gpm) (1,440 min/day) = 3,204,000 gpd

60. 5,350,000 gal/1,440 min/day = 3,715 gpm

61. (2.5 mg/L) (5.5 MGD) (8.34 lb/gal) = 115 lb/day

62. (7.1 mg/L) (4.2 MGD) (8.34 lb/gal) = 249 lb/day

63. (11.8 mg/L) (4.8 MGD) (8.34 lb/gal) = 472 lb/day

64. $$\frac{(10 \text{ mg / L})(1.8 \text{ MGD})(8.34 \text{ lbs / gal})}{0.65} = 231 \text{ lbs/day}$$

65. (41 mg/L) (6.25 MGD) (8.34 lb/gal) = 214 lb/day

66. (60 mg/L) (0.086 MGD) (8.34 lb/gal) = 43 lb

67. (2,220 mg/L) (0.225) (8.34 lb/gal) = 4,166 lb

68. $$\frac{(8 \text{ mg / L})(0.83 \text{ MGD})(8.34 \text{ lbs / gal})}{0.65} = 85 \text{ lbs/day}$$

Answers to More Additional Practice Problems

1. (0.785) (60 ft) (60 ft) (20 ft) (7.48 gal/cu ft) = 422,769.6 gal
2. (50 ft) (15 ft) (10 ft) = 7,500 cu ft
3. (10 ft) (50 ft) (10 ft) (7.48 gal/cu ft) = 37,400 gal
4. (10 ft) (40 ft) (10 ft) (7.48 cu ft) = 29,920 gal
5. (0.785) (50 ft) (50 ft) (12 ft) (7.48 gal/cu ft) = 176,154 gal
6. (20 ft) (40 ft) (16 ft) (7.48 gal/cu ft) = 95,744 gal
7. (4 ft) (6 ft) (300 ft) = 7,200 cu ft
8. (0.785) (0.83 ft) (0.83 ft) (1,500 ft) (7.48 gal/cu ft) = 6,068 gal
9. 5 ft + 10 ft/2 (4 ft) (700 ft) (7.48 gal/cu ft)
   = (7.5 ft) (4 ft) (700 ft) (7.48 gal/cu ft)
   = 157,080 gal
10. (0.785) (0.66) (0.66) (2,100 ft) (7.48 gal/cu ft) = 5,371 gal
11. (5 ft) (4 ft) (1,100 ft) (7.48 gal/cu ft) = 164,560 gal
12. $\dfrac{(4\ ft)(4\ ft)(1100\ ft)}{27\ \text{cu ft/cu yd}} = 652$ cu yds
13. (400 yd) (1 yd) (1.33 yd) = 532 cu yd
14. (810 ft) (3 ft) (3 ft) = 7,290 cu ft
15. (600 ft) (6.5 ft) (3.5 ft) = 13,650 cu ft
16. (0.785) (90 ft) (90 ft) (20 ft) (7.48 gal/cu ft) = 951,232 gal
17. (80 ft) (12 ft) (20 ft) = 19,200 cu ft
18. (0.785) (0.67 ft) (0.67 ft) (3,000 ft) (7.48 gal/cu ft) = 7,908 gal
19. (1,500 ft) (3 ft) (3 ft) = 13,500 cu ft
20. $\dfrac{(3\ ft)(4\ ft)(1100\ ft)}{27\ \text{cu ft/cu yd}} = 489$ cu yds
21. (30 ft) (60 ft) (12 ft) (7.48 gal/cu ft) = 161,568 gal
22. (8 ft) (3.5 ft) (2,000 ft) (7.48 gal/cu ft) = 418,880 gal
23. (0.785) (60 ft) (60 ft) (19 ft) (7.48 gal/cu ft) = 401,631 gal
24. (0.785) (20 ft) (20 ft) (30 ft) (7.48 gal/cu ft) = 70,462 gal
25. (2.4 ft) (3.7 ft) (2.0 fps) (60 sec/min) = 1,066 cfm
26. (20 ft) (12 ft) (0.7 fpm) (7.48 gal/cu ft) = 1,257 gpm
27. $\dfrac{(4\ ft + 6\ ft)}{2}(3.3\ \text{ft})\ (120\ \text{fpm}) = 5(3.3\ \text{ft})\ (120\ \text{fpm})$
    = 1,980 cfm
28. (0.785) (0.66) (0.66) (2.2 fps) (7.48 gal/cu ft) (60 sec/min) = 338 gpm
29. (0.785) (2 ft) (2 ft) (4.7 fpm) (7.48 gal/cu ft) = 110 gpm
30. (0.785) (0.83 ft) (0.83 ft) (3.0 fps) (7.48 gal/cu ft) (60 sec/min) (0.5) = 364 gpm
31. (6 ft) (2.5 ft) (x fps) (60 sec/min) (7.48 gal/cu ft) = 14,200 gpm
    x = 2.11 ft
32. (0.785) (0.67) (0.67) (x fps) (7.48 gal/cu ft) (60 sec/min) = 590 gpm
    x = 3.7 fps

33. (0.785) (0.83 ft) (0.83 ft) (2.4 fps) = (0.785) (0.67 ft) (0.67 ft) (x fps)
    x = 3.7 fps
34. 400 ft/92 sec = 4.3 fps
35. (0.785) (0.67) (0.67) (3.2 fps) = (0.785) (0.83 ft) (0.83 ft) (x fps)
    x = 2.1 fps
36. 35.3 MGD/7 = 5 MGD
37. 124.4 MG/30 days = 4.1 MGD
38. 1,000,000 × 0.175 = 175,000 gpd
39. 3,330,000 gal/1,440 min = 2,313 gpm
40. (7 cfs) (7.48 gal/cu ft) (60 sec/min) = 3,142 gpm
41. (30 gps) (60 sec/min) (1,440 min/day) = 2,592,000 gpd
42. $$\frac{4,500,000 \text{ gpd}}{(1440 \text{ min/day}) \ (7.48 \text{ gal/cu ft})} = 418 \text{ cfm}$$
43. (6.5 MGD) (1.55 cfs/MGD) = 10.1 cfs
44. $$\frac{(445,870 \text{ cfd}) (7.48 \text{ gal / cu ft})}{1440 \text{ min/day}} = 2316 \text{ gpm}$$
45. (2,400 gpm) (1,440 min/day) = 3,456,000 gpd
46. (6 ft) (2.0 ft) (x fps) (7.48 gal/cu ft) (60 sec/min) = 14,800 gpm
    x = 2.7 fps
47. (4.6 ft) (3.4 ft) (3.5 fps) (60 sec/min) = 3,284 cfm
48. 378.6/92 days = 4.1 MGD
49. (10 ft) (10 ft) (0.67 fpm) (7.48 gal/cu ft) = 501 gpm
50. (0.785) (0.67 ft) (0.67 ft) (x fps) (7.48 gal/cu ft) (60 sec/min) = 510 gpm
    x = 3.2 fps
51. (11 cfs) (7.48 gal/cu ft) (60 sec/min) = 4,937 gpm
52. 134.6/30 days = 4.5 MGD
53. (5.0 MGD) (1.55 cfs/MGD) = 7.8 cfs
54. (0.785) (2 ft) (2 ft) (3.3 fpm) (7.48 gal/cu ft) = 77.5 gpm
55. $$\frac{(1,820,000 \text{ gpd})}{(1440 \text{ min/day}) \ (7.48 \text{ gal/cu ft})} = 169 \text{ cfm}$$
56. (0.785) (0.5 ft) (0.5 ft) (2.7 fps) (7.48 gal/cu ft) (60 sec/min) = 238 gpm
57. (0.785) (0.83 ft) (0.83 ft) (2.6 fps) = (0.785) (0.67 ft) (0.67 ft) (x fps) =
    x = 4.0 fps
58. (2,220 gpm) (1,440 min/day) = 3,196,800 gpd
59. 5,300,000 gal/1,440 min/day = 3,681 gpm

Index

Note: Page numbers in **bold** refer to tables.

A

activated carbon, 95
adsorption, 79
aeration, in iron and manganese removal, 30, 32
air
    as oxidant for iron and manganese control, 31
    weight of, related to weight of water, 2
algae control, 15, 27
alkalinity in sedimentation, 65
    total alkalinity required, 65
    see also lime dosage, for sedimentation
alkalinity in water softening, 145–148
    bicarbonate alkalinity, 146–148
    carbonate alkalinity, 146–148
    determination of, 145–146
    hydroxide alkalinity, 146–148
    phenolphthalein alkalinity, 145–146, **147**, 148
    total alkalinity, 145–146, **147**, 148
    in total hardness calculation, 144–145
alum, 35, 38–39, 47–52, 65–67
American Water Works Association (AWWA),
    water fluoridation standards, 126
anthracite, 93, 95
atmospheric pressure, 2
available fluoride ion concentration (AFI), 129, **129**
average daily use
    of chemical, 45–46
    of chlorine, 120–121

B

backwashing, *see* filtration
barium, in water hardness, 141
bicarbonate alkalinity, 146–148
biosolids, volume pumped, 13–14
bleach, 19, 117, 122–123
bowl efficiency, 24
bowl horsepower, 22–23
brake horsepower (BHP), 10
    and pump efficiency, 10–12
    total, 21
breakpoint chlorination, 113–115
brine and salt requirement for regeneration,
    156–158

C

calcium hardness of water, 141–142, 143
calcium hypochlorite, *see* dry hypochlorite

calculated doses in fluoridation process, 133–137
calculations, step-by-step breakdown, 81–82
calibration of chemical feeders, 42–45, 53–54
carbonate alkalinity, 146–148
carbonate hardness of water, 144–145,
    148–150
carbon dioxide dosage, in recarbonation,
    151–152
caustic soda, 126
Centers for Disease Control and Prevention
    (CDC), on sodium fluoride in dry feeders,
    132
centrifugal pumps
    compared with positive displacement
        pumps, 13
    in deep well turbine pump, 19
    head loss calculation, 8–9
    hydraulic horsepower calculation, 10
C factor (pipe wall roughness), 6
chemical feeders, 53–54
    calibration, 42–45, 53–54
    feeder settings, 38–40
    practice problems, 54–57
    types of, 53
chemical usage
    in chlorination, 120–121
    in coagulation and flocculation, 45–46
    in fluoridation, 128–129
    in water softening, 148–151
chloramines, 112, 113
chlorination, 109–124
    average daily use, 120–121
    breakpoint chlorination, 113–115
    chemical inventory, 120–121
    chlorine dosage (feed rate), 109–111
    chlorine dose, demand, and residual, 111–113
    disinfection, 109
    dry hypochlorite feed rate, 115–117
    dry hypochlorite percent strength, 119
    liquid hypochlorite feed rate, 117–118
    liquid hypochlorite percent strength, 120
    percent strength of solutions, 119–120
    practice problems, 121–124
chlorine
    demand, 111–113
    disinfection, 109
    dose, 111–113
    forms of, 109
    gaseous chlorine, 109, 123
    groundwater treatment, 35

as oxidant for iron and manganese control, 31
residual, 109, 111–113
for well casing disinfection, 19
coagulants (salts), 35
coagulation, defined, 35
coagulation and flocculation, 36–46
chambers and basins, volume of, 36–37
chemical usage, 45–46
detention time, 37–38
feeder calibration for dry chemical, 42–43
feeder calibration for solution chemical,
43–45
feeder settings for dry chemical, 38
feeder settings for solution chemical, 39–40
percent of solutions, 40–41
percent strength of liquid solutions, 41–42
percent strength of mixed solutions, 42
practice problems, 46–53
combined residual, of chlorine, 112, 113–114
continuity, law of, 5
conversions
cubic feet/second to gpm or MGD, 4
cubic feet to gallons, 8
gallons/day to gallons/hour, 61
gallons/day to gallons/minute, 79
gallons/day to gallons/second, 38
gallons/minute/ft^2 to inches/minute rise, 87
gallons/minute to gallons/day, 14, 111
gallons/minute to gallons/hour, 18
grams to pounds, 41
head to pressure, 9
inches to feet, 37
kilowatts power to horsepower, 12
milligrams/liter to pounds/day, 38, 68
milliliters/minute to gallons/day, 44
ounces to pounds, 40
pounds/day to gallons/day, 39, 118
pounds/day to grams/minute, 69, 131
pounds/day to pounds/hour, 153
pounds/day to pounds/minute, 152–153
pounds/minute to pounds/day, 43
pressure differential to total differential
head, 10
pressure measurement, 3
pressure to feet of head, 20
copper sulfate dosing, 27–30
Cryptosporidium, 79
cubic feet per second (ft^3/sec or cfs), 4

D

deep well turbine pump, 19–20
deferrization, 30
demand, chlorine, 111–113, 123
demanganization, 30
dental carries, 125, 129
dental fluorosis, 125

detention time
in coagulation and flocculation, 37–38
in sedimentation, 60–61, 76–78
diaphragm pump, 53
discharge head, 20
discharge of water in motion, 4–5
disinfection
chlorination for, 109
of well casing, 19
dosing, defined, 38
drawdown, in wells, 16–17
drip feeder, 53
drop test, 102–103
dry feeder types, 53
dry hypochlorite (calcium hypochlorite)
feed rate, 115–117
percent strength, 119
as solid calcium hypochlorite, 109

E

effective size (ES) of filter medium, 94
efficiency
field, 24–25
filter, 100–101
horsepower, 24
of motor or pump, 11
exchange, *see* ion exchange
expansion of filter bed, 92–93, 97

F

feeder calibration, 42–45, 53–54
feeders, chemical
calculations, 54–57
types of, 53
feeder settings, 38–40
feed rate
chlorine dosage, 109–111, 121–122
in coagulation and flocculation, 42–45
of fluoride, 130–132, 137–139
of fluoride in saturator, 132–133, 137, 139
of hypochlorite, dry calcium, 115–117
of hypochlorite, liquid sodium, 117–118
of lime, in water softening, 152–153
field efficiency, 24–25
field head (total pumping head), 20
field horsepower, 21, 23
filtration, 79–106
about technologies, 79
backwash, percent product water used, 90–91
backwash pumping rate, 89–90
backwash rate, 85–87, 102
backwash rise rate, 87–88
backwash water, volume required, 88
backwash water tank, required depth, 89
filter bed expansion, 92–93

filter classifications, 93
filter efficiency, 100–101
filter loading rate, 93–94, 103
filter medium size, 94–95
filtration rate, 82–84
flow rate through a filter, 79–82
head loss for fixed-bed flow, 96–97
head loss through a fluidized bed, 97–99
horizontal washwater troughs, 99–100
mixed media, 95–96
mud balls, percent volume, 91–92
practice problems, 101–106
unit filter run volume (UFRV), 84–85
fixed-bed flow, head loss for, 96–97
flocculation, defined, 35, *see also* coagulation
and flocculation
flow, mean flow velocity, in sedimentation, 62–63
flow rate through a filter, 79–82
fluidization, 97
fluidized bed, head loss through, 97–99
fluoridation, 125–139
AWWA standards, 126
fluoride compounds, 125–128
fluoride facts, 125
optimal fluoride levels, **128**, 128–129
safety and hazards of fluoride chemicals, 125,
127, 134
theoretical concentration of fluoride, 133–134
in U.S., 125
fluoridation process, 129–139
calculated doses, 133–137
fluoride feed rate, 130–132
fluoride feed rates for saturator, 132–133
percent fluoride ion in compound, 129–130
practice problems, 137–139
fluoride compounds, *see* fluorosilicic acid;
sodium fluoride; sodium fluorosilicate
fluorosilicic acid, **126**, 126–127, **127**, **129**, 131,
135–136, 138–139
foot-pounds/second (ft-lb/sec), 9
free residual, of chlorine, 112, 113
freshwater sources, 15
friction loss, 5–6

G

Galileo number, 98
gallons per hour (gph), 6
gallons per minute (gpm), 4, 8
gas chlorinators, 53
gaseous chlorine, 109, 123
gauge pressure, 3
Giardia lamblia, 79
gluconite, 32
grain size distribution analysis, 94
grains of hardness removal, 153
gravimetric dry feeder, 53

gravity feed rotameter, 53
greensand filter process, 32–34
groundwater, and chlorination, 35

H

hardness of water, 141–145, *see also* softening
water
Hazen–Williams equation, 6
head, 1, 3, 8–9
head loss, 5–6, 8
for fixed-bed flow, 96–97
through a fluidized bed, 97–99
horizontal washwater troughs, 99–100
horsepower, 9
and efficiency, 10–12, 24
in vertical turbine pumps, 21–23
hydrated lime, for carbonate hardness removal,
148–150
hydraulic horsepower (WHP), 9–10
hydraulic loading rate, compared with surface
overflow rate, 61
hydraulics calculations, 1–6
gauge pressure, 3
pipe friction, 5–6
water at rest, 2–3
water in motion, 3–5
weight of water, 1
weight of water related to weight of air, 2
hydrofluoric acid, 127
hydrofluorosilicic acid, *see* fluorosilicic acid
hydroxide alkalinity, 146–148

I

ilmenite sand, 95–96
inventory
of chemical, 45–46
of chlorine, 120–121
ion exchange, in iron and manganese removal,
31–32
ion exchange softening process, 153–158
exchange capacity of resin, 153–154
operating time until regeneration, 156
salt and brine requirement for regeneration,
156–158
salt solutions table, 157
water treatment capacity, until regeneration,
154–155
iron, in water hardness, 141
iron and manganese removal, 30–34
aeration, 32
aesthetic and economic problems, 30
ion exchange, 31–32
oxidation, 31
potassium permanganate oxidation and
manganese greensand, 32–34

precipitation, 30–31
   sequestering, 32
iron bacteria slimes, 30
iron residue in pipes, 30

J

jar tests, 38, 46

K

kinetic pumps, 13
Kozeny equation, 96–97

L

laboratory horsepower, 22
lakes, storage capacity, 25–27
law, Stevin's, 2
law of continuity, 5
lime dosage, for sedimentation
   grams/minute, 69
   milligrams/liter, 65–68
   pounds/day, 68–69
lime-soda ash, water softening process,
   148–153
   feed rates, 152–153
   lime dosage for carbonate hardness removal,
     148–150
   molecular weights of chemicals, 148
   recarbonation, 151–152
   soda ash dosage for noncarbonate hardness
     removal, 150–151
liquid feeder types, 53
liquid hypochlorite (sodium hypochlorite)
   feed rate, 117–118
   percent strength, 120
   as sodium hypochlorite solution, 109
loading rates, for sand filters, 93–94, 103

M

magnesium hardness of water, 142–144
manganese dioxide, 31, 32
manganese greensand, 32
manganese removal, *see* iron and manganese
   removal
math practice problems and answers
   advanced problems, 243–251, 299–300
   area of rectangles, 163, 254
   average flow rates, 239, 298
   channel and pipeline capacity calculations,
     235, 297
   chemical dosage calculations, 242–243, 298
   circumference and area of circles, 163, 254
   decimal operations, 159–160, 253
   find x, 161–162, 253–254
   flow, velocity, and conversion calculations,
     237–239, 297–298
   flow and velocity calculations, 240–242, 298
   flow conversions, 239–240, 298
   percentage calculations, 160, 253
   ratio and proportion, 162, 254
   tank volume calculations, 234, 296–297
   volume calculations, 235–237, 297
   water treatment, 164–234, 254–296
   see also practice problems (with answers)
mean flow velocity, in sedimentation, 62–63
medium size of filters, 94–95
metalic ions, in water hardness, 141
methyl orange alkalinity, 27
microorganisms
   chlorine disinfection of, 109
   iron and manganese in water, 30
milligrams per liter (mg/L), 109, 130
milliliters per minute (mL/min), 39–40
million gallons per day (MGD), 4
minimum fluidizing volume, 97–98
mixed media filter bed, 95–96
motor efficiency, 24
motor horsepower (mhp), 10, 21
mud balls, percent volume, 91–92

N

nitrogen compounds, in chlorination, 112, 113
noncarbonate hardness of water, 144–145, 148,
   150–151

O

operating time, in ion exchange softening, 156
overall efficiency, 24–25
oxidation, in iron and manganese removal, 31

P

parts per million, 130
percent efficiency, 11–12, 24
   of motor, 11–12
   overall, 11
   pump, 11
percent filter bed expansion, 92–93
percent fluoride ion in compound, 129–130
percent mud ball volume, 91–92
percent product water used for backwashing,
   90–91
percent settled biosolids, 64–65
percent strength of solutions
   in chlorination, 119–120
   in coagulation and flocculation, 40–42
perfect vacuum, 2
phenolphthalein alkalinity, 145–146, **147**, 148
pH of water for precipitation, 30–31

phosphate fertilizer, and phosphate rock, 127
pi, 4
pipe
    area of, 4
    equivalent length, 6
    friction, 5–6
    wall roughness (C factor), 6
piston pumps, 13, 53
polymer, 35, 38, 40–41, 44, 47, 49–51
ponds, storage capacity, 25–27
positive displacement pumps, 13–14
    compared with kinetic pumps, 13
    head loss calculation, 8–9
    hydraulic horsepower calculation, 10
    types of solution feeders, 53
    volume of biosolids pumped, 13–14
potassium permanganate oxidation, 31, 32–34
pounds per acre-foot (lb/ac-ft), 27
pounds per day (lb/day), 54, 109
pounds per square foot (lb/ft²), 3
pounds per square inch (lb/in²), 3
power, 9
practice problems (with answers)
    chemical feeders, 54–57
    chlorination, 121–124
    coagulation and flocculation, 46–53
    filtration, 101–106
    fluoridation, 137–139
    sedimentation, 70–78
    see also math practice problems and answers
precipitation, in iron and manganese removal, 30–31
pressure, air and water, 1–3
pump column, 19–20
pump efficiency and brake horsepower (BHP), 10–12
pumping, 1
pumping calculations, 6–14
    head calculation, 8–9
    head loss calculation, 8
    horsepower and efficiency, 10–12
    hydraulic horsepower (WHP), 10
    pump efficiency and brake horsepower (BHP), 10–12
    pumping rates, 6–8
    specific speed, 12–13
pumping rates, for backwash, 89–90
pumping water level, 16–17
pumps
    deep well turbine pump, 19–20
    diaphragm pump, 53
    gravity feed rotameter, 53
    kinetic pumps, 13
    piston pumps, 13, 53
    reciprocating action pumps, 13, 53
    rotary action pumps, 13
    submersible-type pump, 20

    vacuum type, 53
    vertical turbine pumps, 20–25
    see also centrifugal pumps; positive displacement pumps

Q

quicklime, for carbonate hardness removal, 148–149

R

rapid sand filters, 93–94, 100
recarbonation, in water softening, 151–152
reciprocating action pumps, 13, 53
regeneration, *see* ion exchange softening process
residual, in chlorination, 109, 111–113
resin, in ion exchange, 153, *see also* ion exchange softening process
rotary action pumps, 13

S

safety and hazards of fluoride chemicals, 125, 127, 134
salt
    and brine requirement for regeneration, 156–158
    dissolved in water, 119
salts (coagulants), 35
salt solutions table, 157
sand filters, 80, 93
secondary maximum contaminant levels (SMCLs), 30
sedimentation, 59–78
    defined, 59, 79
    detention time, 60–61
    lime dosage (g/min), 69
    lime dosage (lb/day), 68–69
    lime dosage (mg/L), 65–68
    mean flow velocity, 62–63
    percent settle biosolids, 64–65
    practice problems, 70–78
    surface overflow rate (SOR), 61–62
    tank volume, 59–60
    weir loading rate, 63–64
sequestering, in iron and manganese removal, 32
sieve analysis curve, 94–95
silica sand, 95
silicofluoric acid, *see* fluorosilicic acid
slow sand filters, 80, 93
slurry samples, 64–65
small water systems
    filtration technologies used, 79
    groundwater chlorination, 35

soda ash dosage for noncarbonate hardness
    removal, 150–151
sodium chloride, 156–157
sodium fluoride, 126, **126**, **129**, 132,
    136–137
sodium fluorosilicate, **126**, 127–128, **129**, 131,
    135, 138–139
sodium hypochlorite, *see* liquid hypochlorite
softening water, 141–158
    alkalinity in water, 145–148
    ion exchange softening process, 153–158
    lime-soda ash process, 148–153
    water hardness, 141–145
    see also alkalinity in water softening; ion
        exchange softening process; lime-soda
        ash; water hardness
solute, 119
solution, 119
solution feeder types, 53
solvent, 119
sounding line, 17
specific gravity, 10, 95
specific speed, 12–13
specific yield, 18–19
static water level, 16–17
Stevin's law, 2
straining, 79
strontium, in water hardness, 141
submersible-type pump, 20
sulfuric acid, 126, 127, 146
surface overflow rate (SOR, surface loading rate,
        surface settling rate)
    compared with hydraulic loading rate, 61
    in sedimentation, 61–62, 70–72
Surface Water Treatment Rule (SWTR), 79

T

tanks
    for backwash water, required depth, 89
    for sedimentation, volume of, 59–60
theoretical concentration of fluoride,
        133–134
total alkalinity
    in sedimentation, 65
    in water softening, 145–146, **147**, 148
total brake horsepower, 21
total chlorine residual, 112, 122
total differential head (TDH), 10
total hardness of water, 143–145
total pumping head (field head), 20
tuberculation (iron residue), 30

U

uniformity coefficient (UC), 94–95
unit filter run volume (UFRV), 84–85

V

vacuum, 2
vacuum pump solution feeder, 53
velocity
    mean flow velocity, in sedimentation,
        62–63
    of water in pipe, 4–5
vertical turbine pump, 20–25
volume
    of backwash water required, 88
    of biosolids pumped, 13–14
    of chambers and basins for coagulation and
        flocculation, 36–37
    percent mud ball, 91–92
    of pond or lake, 25–27
    pumped in cubic feet, 8
    of tanks for sedimentation (rectangular and
        circular), 59–60
volume over volume test (V/V test), 64
volumetric dry feeder, 53

W

washwater troughs, 99–100
wastewater troughs, 99
water
    absolute viscosity, 98
    density, 3
    discharge, 4–5
    groundwater and chlorination, 35
    in motion, 3–5
    at rest, 2–3
    sources of freshwater, 15
    storage, 25–27
    upward flow (fluidization), 97
    weight of, 1
    weight of water related to weight of air, 2
    see also chlorination; coagulation and
        flocculation; filtration; fluoridation;
        sedimentation; softening water
water hardness, 141–145
    calcium hardness, 141–142
    carbonate hardness, 144–145
    magnesium hardness, 142–143
    noncarbonate hardness, 144–145
    total hardness, 143–145
    see also softening water
water horsepower, 10, 22–23
water pressure, 1, 3
water source calculations, 16–25
    deep well turbine pump, 19–20
    freshwater sources, 15
    specific yield, 18–19
    vertical turbine pump calculations,
        20–25
    well casing disinfection, 19

well drawdown, 16–17
   well yield, 17–18
water storage, 25–27
weir loading rate (weir overflow rate, WOR), in
   sedimentation, 63–64, 74–75
well casing disinfection, 19

well drawdown, 16–17
well yield, 17–18

Z

zeolite, 32

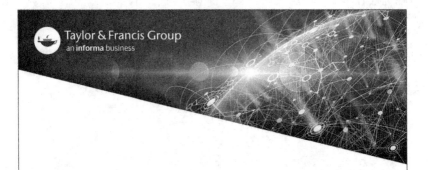

Printed in the United States
by Baker & Taylor Publisher Services